FLORA ZAMBESIACA

Flora terrarum Zambesii aquis conjunctarum

VOLUME NINE: PART FIVE

FLORA ZAMBESIACA

MOZAMBIQUE

MALAWI, ZAMBIA, ZIMBABWE

BOTSWANA

VOLUME NINE: PART FIVE

Edited by
G.V. POPE

on behalf of the Editorial Board:

S.J. OWENS
Royal Botanic Gardens, Kew

M.A. DINIZ
*Centro de Botânica, Instituto de Investigação
Científica Tropical, Lisboa*

G.V. POPE
Royal Botanic Gardens, Kew

Published by the Royal Botanic Gardens, Kew,
for the Flora Zambesiaca Managing Committee
2001

Typesetting and page make-up by Media Resources, Information Services Department,
Royal Botanic Gardens, Kew

Printed in the European Union by
The Cromwell Press

ISBN 1 84246 040 4

CONTENTS

The family Euphorbiaceae is published in two parts with continuous pagination. Vol. 9 part 4 (pages 1–337) deals with subfamilies Phyllanthoideae, Oldfieldioideae, Acalyphoideae, Crotonoideae and Euphorbioideae (tribe Hippomaneae). Vol. 9 part 5 (pages 339–465), with subfamily Euphorbioideae tribe Euphorbieae, completes the Flora account.

INCLUDED IN VOLUME IX, PART 5

153. Euphorbiaceae
Subfamily: Euphorbioideae tribe Euphorbieae

No new names are published in this part

Correction to vol. 9 part 4

On page 242, the sentence starting on line 21 from the bottom should read:
Female flowers: calyx as in the male; petals absent; disk minute or absent; ovary
ovoid, 3(4)-locular, with 1 ovule per locule; styles short, stigmas thick, spreading,
entire or slightly 2-lobed.

Acknowledgements

The Flora Zambesiaca Managing Committee thanks M.A. Diniz and E. Martins of
the Centro de Botânica, Lisbon, for their valuable help in reading and
commenting on the text.

Tribe EUPHORBIEAE

by S. Carter & L.C. Leach

Tribe **Euphorbieae** Blume, Bijdr. Fl. Ned. Ind.: 631 (1826). —Radcliffe-Smith, Gen. Euphorbiacearum: 398 (2001).

Trees, shrubs, or herbs, often succulent and sometimes spiny, always with a milky latex, monoecious or occasionally dioecious. Leaves alternate or verticillate, or sometimes opposite, simple, entire or toothed. Stipules present and often modified as glands or spines, or absent. Inflorescence terminal or axillary, often (pseudo)umbellate, composed of cyathia arranged in simple or dichotomously branching cymes. Bracts paired. Cyathia with numerous male flowers in 5 groups surrounding a solitary female flower and all enclosed within a cup-like involucre (TAB. **71**, figs. 1–4). Involucre with a glandular rim surrounding 5 toothed lobes. Each male flower reduced to a single articulated pedicellate stamen without a perianth, bracteolate. Female flower consisting of a pedicellate pistil; perianth absent or reduced to a rim, occasionally 3-lobed. Ovary 3-celled, with 1 pendulous ovule in each cell; styles 3, usually united at the base; stigmas usually bifid. Fruit usually 3-lobed, usually a capsule dehiscing loculicidally and septicidally (regma) (TAB. **71**, figs. 6 & 7), rarely indehiscent and drupaceous. Seeds with or without a caruncle.

A well-defined tribe united by the uniform nature of the cyathium, comprising the very large genus *Euphorbia*, itself with a world-wide distribution, and 10 smaller genera in tropical and subtropical regions. Of these 6 are endemic in tropical Africa, the 2 largest occurring in the Flora Zambesiaca area (*Synadenium* and *Monadenium*). *Pedilanthus*, from tropical America, is sometimes cultivated.

1. Involucral glands completely enclosed by the cyathium · · · · · · · · · · · · **62. Pedilanthus**
– Involucral glands exposed round the rim of the involucre · 2
2. Involucral glands distinct (TAB. **71**, figs. 8–11), 4–8 or single and funnel-shaped · · · · · · ·
· **59. Euphorbia***
– Involucral glands forming a continuous rim, circular or with a gap to one side · · · · · · · 3
3. Glandular rim circular, sometimes cleft but not widely so · · · · · · · · · · · · **60. Synadenium**
– Glandular rim horseshoe-shaped, with a wide gap to one side · · · · · · · · **61. Monadenium**

59. EUPHORBIA L.

Euphorbia L., Sp. Pl.: 450 (1753); Gen. Pl., ed. 5: 208 (1754). —Boissier in de Candolle, Prodr. **15**, 2: 7 (1862). —Bentham & Hooker, Gen. Pl. **3**, 1: 258 (1880). —N.E. Brown in F.T.A. **6**, 1: 470 (1911); F.C. **5**, 2: 222 (1915). —Pax & K. Hoffmann in Engler, Nat. Pflanzenfam., ed. 2, **19c**: 208 (1931). —S. Carter in F.T.E.A., Euphorbiaceae, part 2: 409 (1988). —Radcliffe-Smith, Gen. Euphorbiacearum: 405 (2001).

Herbs, shrubs or trees, sometimes succulent and unarmed or spiny, with a milky usually caustic latex, monoecious, rarely dioecious. Roots fibrous, or thick and fleshy, sometimes tuberous. Leaves opposite, alternate or verticillate, often stipulate, subtended by spiny outgrowths in some succulent species. Inflorescence with cyathia in simple, dichotomous or umbellate terminal or axillary cymes; bracts paired, often leaf-like. Involucres with (1)4–5(8) nectiferous glands around the rim alternating with 5 fringed lobes. Male flowers bracteolate, in 5 groups separated by fringed membranes. Female flower subsessile or with the pedicel elongating and

---

* Section *Chamaesyce*, as it occurs in tropical Africa, would be particularly easy to distinguish as a genus, and since this account was submitted for publication the evidence for doing so has been accepted by almost all workers.

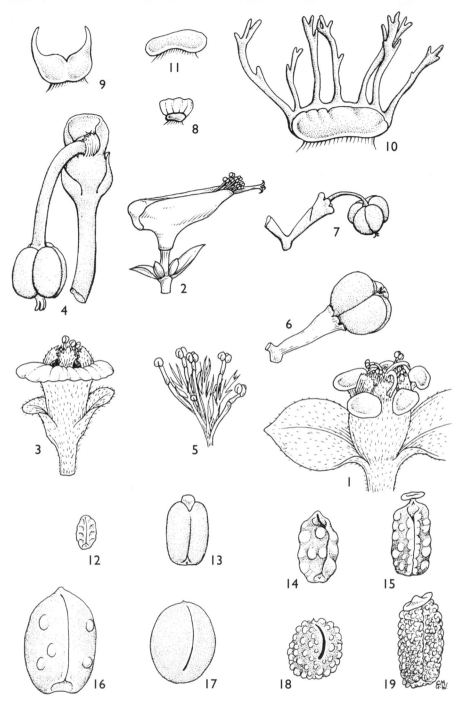

Tab. 71. Key characters of Euphorbieae. **Cyathia**: 1, *Euphorbia* with glands distinct; 2, *Pedilanthus* with glands enclosed; 3, *Synadenium* with glands forming a complete circle; 4, *Monadenium* with glandular rim open at one side. **Stamens and bracteoles**: 5, *Euphorbia*. **Capsules**: 6, sessile or subsessile; 7, exserted. **Involucral glands**: 8, *Euphorbia* sect. *Chamaesyce*; 9, sect. *Esula*; 10, sect. *Trichadenia*; 11, sect. *Euphorbia*. **Seeds**: 12, *Euphorbia* sect. *Chamaesyce*; 13, sect. *Esula*; 14, sect. *Pseudacalypha*; 15, sect. *Eremophyton*; 16, sect. *Trichadenia*; 17, sect. *Lyciopsis*; 18, sect. *Euphorbia*; 19, *Monadenium*. Drawn by Christine Grey-Wilson. From F.T.E.A.

reflexed in fruit but straightening before dehiscence; perianth reduced to a rim below the ovary, rarely 3-lobed; styles partly united. Capsule sometimes fleshy but becoming woody at maturity, dehiscent. Seeds with or without a caruncle.

Euphorbia is one of the largest genera of flowering plants, with possibly over 2000 species (or at least distinct taxa). Of these, about 1300 are herbaceous with a world-wide distribution in temperate and tropical zones. The other species are shrubs, trees or succulents, confined almost entirely to the tropics and subtropics. With relatively few exceptions the succulent species, numbering over 500, occur in drier regions of the African continent and Madagascar, and are well adapted to survive in often extreme xerophytic conditions. Their growth habit is extremely diverse from small herbs to shrubs and large trees.

The latex of all species is abundant and usually caustic, sometimes exceedingly so. Because of this, and its irritant qualities, it is often a component, in Africa, of arrow and fish poisons, and is used widely in native medicines. As such it is known occasionally to promote some forms of cancer and consequently its chemical properties are of importance in cancer research. The latex from some larger succulent species is still used locally in bird-lime. Attempts have been made in the past to use it as a rubber substitute, but the resulting compounds proved to be unstable. Currently some species are being investigated for their hydrocarbon content as a source of fuel, but so far its extraction has been uneconomic.

Because of the unifying structure of the cyathium, classification of this large and otherwise diverse aggregate of species has always been difficult and no system proposed so far has proved entirely satisfactory. Some authors have separated *Euphorbia* into a number of genera, with reasons usually based upon peculiarities of the cyathial structure. As systematic treatments are often confined to limited geographical areas, such concepts invariably break down when applied to related species from elsewhere. Section *Chamaesyce*, as it occurs in tropical Africa, would be particularly easy to distinguish as a genus, and since this account was submitted for publication the evidence for doing so is now accepted by almost all workers. However, Section *Diacanthium* of Boissier and Pax, accommodating all species with succulent stems bearing paired spines, is even more precisely defined. To follow a uniform treatment, this last section would have to be separated as *Euphorbia* sensu stricto (since it contains the type of the genus, namely *E. antiquorum* L.), leaving a large heterogeneous group of species in Africa and elsewhere, often with extremely ill-defined generic limits. Without a study of *Euphorbia* as a whole, and also to follow the treatment in Flora Tropical East Africa (1988), it seems advisable to refrain from separation at generic level in the Flora Zambesiaca account.

The systematic arrangement employed here loosely follows Pax's adaptation (1931) of Boissier's work, with modifications to better accommodate a number of more-or-less distinct groups recognised within the Flora Zambesiaca area and in East Africa. These are placed in Sections in a sequence which is not strictly phylogenetic but reflects increasing specialization towards succulence.

SYNOPSIS OF THE SECTIONS OF *EUPHORBIA* IN THE FLORA ZAMBESIACA AREA

Sect. 1. HYPERICIFOLIAE Pojero. Herbaceous annuals with an abbreviated main stem, opposite leaves, obvious stipules, 4 involucral glands with petaloid appendages on the outer margin, and ecarunculate seeds. Includes species with 10 or more cyathia grouped or congested into often capitate cymes with bracts much reduced. Species 2, introduced and common throughout the Flora Zambesiaca area. *Species 1, 2.*

Sect. 2. CHAMAESYCE (S.F. Gray) Reichb. This Section and Section *Hypericifoliae* are regarded by many authors as belonging to an easily defined genus, *Chamaesyce* S.F. Gray (Nat. Br. Pl. **2**: 260 (1821)), which is principally of New World origin and possesses several distinguishing and often advanced characters. Most species are small herbaceous annuals. The main stem aborts at the seedling stage, and as a result the entire plant at maturity consists of an expanded dichotomously branching pseudumbellate inflorescence, with the floral bracts assuming the appearance and function of normal leaves. Stipules are obvious, the 4 involucral glands bear petaloid appendages on the outer margin, and the seeds are ecarunculate. This Section includes species with solitary cyathia, or a few clustered in leafy cymes. Three species in the Flora Zambesiaca area are obvious perennials. *Species 3–20.*

Sect. 3. ARTHROTHAMNUS (Klotzsch & Garcke) Boiss. Characterised by several features present in the previous Sections. Shrubs with a much reduced main stem, cylindrical succulent branches, opposite leaves (actually bracts of the inflorescence) much reduced and quickly deciduous, stipules usually absent, terminal cymes branching dichotomously many times, tiny cyathia with 5 glands, subsessile capsules, conical obtusely 4-angled ecarunculate seeds. South African, with one species just extending into Botswana. *Species 21.*

Sect. 4. POINSETTIA (Graham) Baill. Of New World origin and often regarded as a separate genus. The floral bracts are large and often decorative, and the single involucral gland is distinctively funnel-shaped. Two herbaceous species have been introduced and become naturalised. The large ornamental shrub, *Poinsettia* itself (*Euphorbia pulcherrima*) is commonly cultivated and is recorded as an established escape in Malawi on the lower slopes of the Zomba Plateau. *Species 22–24.*

Sect. 5. PSEUDACALYPHA Boiss. Exclusively African annual or short-lived perennial herbs, or woody-based shrubs, with petiolate leaves, stipules modified as glands or subulate filaments, leafy bracts and usually 4 involucral glands. Used in its broadest sense, this Section includes species with axillary and also usually terminal simple or umbellate cymes, a subglobose capsule with truncate base and ecarunculate seeds with horizontal ridges. In its narrowest sense it is typified by *Euphorbia acalyphoides* Boiss. to include only those species with axillary cymes which dichotomize just once or twice, and involucral glands hairy on the upper surface. There are no representatives in the Flora Zambesiaca area. Other species are then placed in Section *Holstianae* Pax & K. Hoffm. (in Engler, Pflanzenw. Afrikas [Veg. Erde 9] **3**, 2: 148 (1921)), characterised by axillary and terminal 3-rayed umbels branching dichotomously many times, and glabrous involucral glands. The species in Flora Zambesiaca belong here. *Species 25–27.*

Sect. 6. EREMOPHYTON Boiss. Includes woody-stemmed herbaceous annual or short-lived perennial species with petiolate leaves and vestigial stipules, terminal and axillary 3-rayed umbels branching dichotomously, elongated capsules and flattened seeds with a conspicuous cap-like caruncle. Almost all species are African, of which one species only occurs in the Flora Zambesiaca area, in Botswana and southern Zimbabwe. *Species 28.*

Sect. 7. ESULA Dumort. One of several very closely related Sections which are generally accepted as the most primitive in the genus, although species possess some characters more advanced than those of many species from other sections occurring in Africa, such as the development of an herbaceous habit, loss of stipules and a condensed pseudumbellate inflorescence. The group has diversified, especially in temperate zones, to include the largest number of species. All those in the Flora Zambesiaca area belong to Section *Esula*, distinguished by leaf-like bracts, 4 involucral glands and carunculate seeds. All are indigenous except for *Euphorbia peplus. Species 29–39.*

Sect. 8. TRICHADENIA Pax & K. Hoffm. Includes herbs, and trees and shrubs with woody or semi-succulent branches, thick, fleshy, often tuberous roots, dichotomously branching 3–many-branched pseudumbels, leaf-like bracts, stipules modified as glands or subulate filaments, 4 or 5 crenulate, lobed or large pectinate involucral glands and relatively large capsules with ecarunculate seeds. All species are African. *Species 40–50.*

Sect. 9. PSEUDEUPHORBIUM (Pax) Pax & K. Hoffm. This Section was erected to accommodate *Euphorbia monteiri*. General characters are similar to those of Section *Trichadenia*, but the fleshy tuberous roots are lacking and the main stem is thick and succulent with prominent tubercles. *Species 51, 52.*

Sect. 10. MEDUSEA (Haw.) Pax & K. Hoffm. Herbs with a usually large fleshy root crowned by a truncated stem producing densely packed succulent tuberculate branches. Leaves small, caducous. Involucral glands variously shaped and ornamented. All representatives are South African, with only two species extending across the southern boundaries of Botswana and Zimbabwe. *Species 53, 54.*

Sect. 11. LYCIOPSIS Boiss. A well-defined Section of Old World species including woody-based perennial herbs with a tuberous rootstock, or woody shrubs or small trees, with leaf-like or scarious bracts, conspicuous glandular stipules, terminal or axillary cymes in umbels of up to 7 lateral branches, cyathia with 5 often saucer-shaped glands, sessile capsules and subglobose ecarunculate seeds. *Species 55–57.*

Sect. 12. ESPINOSAE Pax & K. Hoffm. Predominantly South African, with only *Euphorbia espinosa* Pax extending northwards beyond Botswana and Zimbabwe as far as East Africa. Included are woody shrubs or small trees with prominent glandular stipules, scarious bracts and solitary cyathia with 5 involucral glands. These features are characteristic of the previous Section, but the large deeply lobed exserted capsule and smooth seeds with a conspicuous caruncle are typical features of Section *Tirucalli* which follows. *Species 58, 59.*

Sect. 13. TIRUCALLI Boiss. Characteristic features include cylindrical succulent branches, with small quickly deciduous leaves, exserted capsules and carunculate seeds. Some species that have proliferated in Madagascar and southwestern Africa possess glandular stipules, an often congested inflorescence of a pseudumbel with dichotomously branching rays, small scarious bracts and 5 involucral glands. Leaf scars become conspicuously calloused in others that lack stipules, cymes are in terminal pseudumbels of 3–8 rays, and bracts are leafy. Most of these latter species have developed in east and northeast Africa (including the Arabian Peninsula), with some in southwest Africa. *Species 60, 61.*

Sect. 14. EUPHORBIA Includes all those species from the Old World, mostly from Africa, with succulent stems and branches, and stipules which have become modified as prickles mounted on a horny pad (the spine shield) which surrounds or subtends the leaf scar and bears in addition a pair of spines. Stems and branches often with projecting longitudinal angles (angular); angles sometimes thick and fleshy, or thin and wing-like, bearing the spine shields which are sometimes joined to form a horny margin. Inflorescences are always axillary, consisting of one or more cymes branching usually once only, a feature which probably represents a reduction of a more complex pseudumbellate arrangement. Involucral glands number 5 and are always entire. Capsules may be sessile or exserted, a feature indicative of species relationships. Seeds are subglobose, smooth or verrucose (with warts), and ecarunculate.

The largest number of species in tropical Africa are included in this Section, which would be better regarded as a Subgenus, with several Sections based on position, shape and size of the capsule, seed characters, development of the inflorescence, spinescence structure and stem-angles. However, such Sections have not yet been satisfactorily defined. The tree habit has persisted in some species, while others show extreme specialization, involving increased succulence of the stems and branches, and a reduction in size. *Species 62–119.*

CULTIVATED SPECIES

Key to the species of Euphorbia introduced and cultivated in the Flora Zambesiaca area

1. Stems and branches unarmed ·· 2
– Stems and branches spiny ·· 8
2. Involucral glands with petaloid appendages ························· 3
– Involucral glands without appendages ····························· 5
3. Petaloid appendages large, scarlet ···················· iii) *Euphorbia fulgens*
– Petaloid appendages inconspicuous ······························· 4
4. Leaves purple; bracts small ·························· i) *Euphorbia cotinifolia*
– Leaves green; bracts showy, white ··················· ii) *Euphorbia leucocephala*
5. Bracts green; involucral glands 4–5 ······························· 6
– Bracts red, yellow, or blotched; involucral glands 1 ················· 7
6. Stems and leaves hairy ··························· vi) *Euphorbia villosa*
– Stems and leaves glabrous ······················ vii) *Euphorbia lathyris*
7. Herb to 1 m tall; bracts with a conspicuous red blotch at the base ·· iv) *Euphorbia cyathophora*
– Robust shrub to 4 m tall; bracts brilliant red, occasionally pink or yellow ············
  ·· v) *Euphorbia pulcherrima*
8. Stems sprawling with spreading branches, c. 1 cm thick; leaves to c. 6 × 2 cm, ovate ·····
  ··· viii) *Euphorbia milii* var. *splendens*
– Stems erect, sparsely branched, c. 2 cm thick; leaves to c. 12 × 3 cm, lanceolate ········
  ··· viii) *Euphorbia milii* var. *hislopii*

A number of indigenous species, mostly spiny, which are popularly cultivated as garden ornamentals, are not included here, e.g. *E. ingens, E. cooperi, E. griseola, E. tirucalli*.

i) *Euphorbia cotinifolia* L., Sp. Pl.: 453 (1753). This popular garden ornamental, a small shrubby tree originating from central America, produces vast numbers of seedlings and reproduces easily. It grows to c. 5 m high and is prized for its ornamental foliage, its leaves being purplish-brown and broadly ovate with long petioles. The inflorescences are relatively inconspicuous, produced as small, few-branched cymes consisting of tiny cyathia bearing purplish glands with creamy lobed petaloid appendages.

Zimbabwe. C: Harare, Elizabeth Nursery, fl. 30.xi.1974, *Biegel* 4706 (K; SRGH).

ii) *Euphorbia leucocephala* Lotsy in Coult. Bot. Gaz. **20**: 350 (1895). A densely branching shrub to c. 2 m high, introduced from Central America as an ornamental. Its small lanceolate leaves c. 3 cm long are unexceptional, but in flower the branch apices are smothered with clusters of tiny cyathia bearing inconspicuous petaloid glands and subtended by showy creamy-white oblanceolate bracts c. 1 cm long.

Zimbabwe. C: Harare, Landscapes Nurseries, fr. 18.vii.1974, *Biegel* 4519 (K; SRGH).

iii) *Euphorbia fulgens* Karw. ex Klotzsch in Allg. Gartenzeitung **2**: 26 (1834). A small shrub with arching branches, and cyathia c. 1 cm in diameter with yellow glands bearing large bright red petaloid appendages. A native of Central America, it is difficult to propagate and not widely grown.

Zambia. W: Kitwe, cultivated in gardens, fl. 8.x.1968, *Fanshawe* (SRGH). **Zimbabwe**. C: Harare, Landscapes Nurseries, fl. 11.x.1972, *Biegel* 4004 (SRGH).

iv) *Euphorbia cyathophora* Murray in Comment. Soc. Regiae Sci. Gött. **7**: 81, t. 1 (1786). See s*pecies 23*. An annual herb, introduced from Central America, grown for its ornamental bracts, which are markedly panduriform with a bright orange-red basal blotch. Its cyathia have a single funnel-shaped gland.

Zimbabwe. C: Harare, Marlborough Nursery, fl. & fr. 8.iv.1975, *Biegel* 4959 (SRGH).

v) *Euphorbia pulcherrima* Willd. ex Klotzsch in Allg. Gartenzeitung **2**: 27 (1834). —White, F.F.N.R.: 199 (1962). See s*pecies 24*. Usually called *Poinsettia*, this species from Central America is commonly grown as an ornamental shrub. It usually grows to c. 3 m high but can reach c. 4 m high, and is characterised by the large (to c. 15 cm long) brilliant red, occasionally pink or creamy-yellow bracts of the inflorescence. Its cyathia have a single bright yellow funnel-shaped gland.

Zambia. C: Lusaka Forest Nursery, fl. 3.iii.1952, *White* 2168 (K). **Zimbabwe**. C: Harare, Landscapes Nurseries, st. 1.viii.1974, *Biegel* 4538 (SRGH). **Malawi**. C: Dedza, Sylvicultural Arboretum, fl. 1.iv.1968, *Salubeni* 1034 (SRGH). **Mozambique**. M: Maputo, fl. 20.vi.1973, *Balsinhas* 2520 (K; SRGH).

vi) *Euphorbia villosa* Willd., Sp. Pl. **2**: 909 (1800). Densely tufted perennial herb up to 1 m high, with stems glabrous to pilose. The leaves are minutely toothed towards the apex, the bracts are yellowish, and the capsules are tuberculate and glabrous to densely villous. It is a European species, best suited to damp shady places.

Zimbabwe. C: Harare, Glen Forest, fl. 8.x.1982, *Biegel* 5914 (K; MAL; SRGH).

vii) *Euphorbia lathyris* L., Sp. Pl.: 457 (1753). Biennial herb to 1.5 m high, glabrous and glaucous. The oblong-lanceolate entire leaves are conspicuously decussate, and the 2–4 rays of the spreading umbellate inflorescence are up to 8 times dichotomous. This is the Caper Spurge, a weed of cultivated ground in Europe, so called because of the hot peppery tasting latex in the caper-like seeds, which should not be consumed.

Zimbabwe. C: Harare, fr. 19.i.1979, *Wallis* s.n. in *GHS* 263236 (SRGH).

viii) *Euphorbia milii* Des Moul. in Bull. Hist. Nat. Soc. Linn. Bordeaux **1**: 27 (1826). Very spiny semi-succulent-stemmed species native to Madagascar, with a purplish-brown bark, dark green ovate to lanceolate leaves 1–5 cm long, and with paired subcircular brilliant red, or occasionally yellow petaloid bracts (cyathophylls) below the sessile cyathia. There are a number of varieties of which var. *splendens*, often known as the "Christ Thorn", or "Crown of Thorns", is most popularly cultivated. It is semi-prostrate and quickly forms a low dense hedge which can be easily trimmed without harming the plant. Var. *hislopii* is another popular form and is more sturdy with an erect habit.

Var. *splendens* (Hook.) Ursch & Leandri in Mém. Inst. Sc. Madagascar, Sér. B, Biol. Vég. **5**: 148 (1954) [*E. splendens* Hook. —White, F.F.N.R.: 199 (1962)]. Stems sprawling with spreading branches, c. 1 cm thick; leaves to c. 6 × 2 cm, ovate.

Zambia. C: Kamaila Forest, 40 km north of Lusaka, cult., fl. 12.ix.1963, *Angus* 3728 (K). **Zimbabwe**. C: Harare, National Botanic Garden, fl. 10.vii.1974, *Biegel* 4511 (K; SRGH). **Malawi**. C: Dedza, Chongoni Arboretum Garden, fl. 11.xi.1972, *Salubeni* 1864 (MAL; SRGH). **Mozambique**. M: Namaacha, fl. 10.i.1948, *Torre* 7098 (LISC).

Var. *hislopii* (N.E. Br.) Ursch & Leandri in Mém. Inst. Sc. Madagascar, Sér. B, Biol. Vég. **5**: 148 (1954). Stems erect, sparsely branched, c. 2 cm thick; leaves to c. 12 × 3 cm, lanceolate.

Zimbabwe. C: Harare, National Botanic Garden, fl. 10.vii.1974, *Biegel* 4510 (K; SRGH).

Key to the Sections of Euphorbia

1. Herbs; leaves opposite; involucral glands with petaloid appendages · · · · · · · · · · · · · · 2
 – Herbs, shrubs or trees; leaves alternate, or if opposite then minute and deciduous; involucral glands without petaloid appendages · 3
2. Cyathia 10 or more in capitate cymes with small bracts ·
 · Sect. 1. HYPERICIFOLIAE (page 345)

Sect. 1. HYPERICIFOLIAE Pojero, Fl. Sicula 2, **3**: 327 (1907). —S. Carter in F.T.E.A., Euphorbiaceae, part 2: 415 (1988).

Annual herbs, erect or prostrate, branching from near the base. Leaves (bracts) opposite, the base obliquely rounded; stipules present. Cyathia clustered, 10 or more together in terminal and axillary usually pedunculate capitate cymes. Involucres bisexual with 4 glands; glands with an entire petaloid appendage on the outer margin. Stamens just exserted, with subsessile anthers. Perianth of the female flower reduced to a rim below the ovary; styles joined at the base only. Seeds oblong-conical, obtusely 4-angled, without a caruncle.

Pubescence on stems, leaves and capsules of white hairs interspersed with long yellow hairs · 1. *hirta*
Pubescence of white hairs only · 2. *indica*

1. **Euphorbia hirta** L., Sp. Pl.: 454 (1753). —N.E. Brown in F.T.A. **6**, 1: 496 (1911); in F.C. **5**, 2: 249 (1915). —Keay in F.W.T.A., ed. 2, **1**: 419 (1958). —Hadidi in Bull. Jard. Bot. Belg. **43**: 86 (1973). —Agnew, Upland Kenya Wild Fl.: 221 (1974). —S. Carter in Kew Bull. **39**: 643 (1984); in F.T.E.A., Euphorbiaceae, part 2: 415 (1988). Syntypes from India.
   Euphorbia pilulifera sensu Boissier in de Candolle, Prodr. **15**, 2: 21 (1862) non L., see N.E. Brown in F.T.A. **6**, 1: 496 (1911).
   Euphorbia pilulifera var. *procumbens* Boiss. in de Candolle, Prodr. **15**, 2: 21 (1862). Syntypes from tropical America and Asia.
   Euphorbia hirta var. *procumbens* (Boiss.) N.E. Br. in F.T.A. **6**, 1: 497 (1911). Syntypes as above.

Annual herb, prostrate to ascending, with branches to 50 cm long; whole plant pilose, including inflorescence and capsules, with minute white appressed hairs interspersed by yellow spreading segmented hairs c. 1.5 mm long principally on the branches and especially on younger growth. Leaves with a petiole to 3.5 mm long; stipules to 2.5 mm long, linear, rarely laciniate on lush specimens; lamina 1–4 × 0.5–2 cm, ovate, apex subacute, base very obliquely rounded, margin finely toothed, upper surface sometimes almost glabrous, often blotched with purple especially in the region of the midrib. Cymes terminal and axillary, densely capitate to 15 mm diameter on peduncles to 15(20) mm long, occasionally subtended by 1–2 leaf-like bracts c. 1 cm long; cyathial peduncles to 1 mm long. Bracts to 1 mm long, deltoid, deeply laciniate. Cyathia c. 0.8 × 0.8 mm, with cup-shaped involucres, usually tinged purple; glands 4, minute, elliptical, green or purplish, with minute entire white to pink appendages; lobes triangular, fimbriate. Male flowers: bracteoles linear, fimbriate; stamens 1 mm long. Female flower: ovary shortly pedicellate; styles 0.4 mm long, spreading, bifid almost to the base, with thickened apices. Capsule just exserted on a pedicel 1 mm long, 1 × 1.25 mm, acutely 3-lobed with truncate base, pilose with short yellow appressed hairs. Seeds 0.8 × 0.4 mm, oblong-conical with obscure transverse wrinkles, pinkish-brown.

Botswana. N: Maun, fl. & fr. 30.xi.1974, *P.A. Smith* 1213 (K; SRGH). **Zambia**. B: Kalabo Resthouse, fl. & fr. 13.xi.1959, *Drummond & Cookson* 6412 (K; SRGH). W: Ndola, fl. & fr. 13.ii.1954, *Fanshawe* 806 (K; NDO). C: South Luangwa National Park, Mfuwe, fl. & fr. 12.xii.1968, *Astle* 5390 (K; NDO; SRGH). E: Chipata Distr., Luangwa Valley, Mulila Munkanya, 2.iii.1968, *R. Phiri* 78 (K; NDO; SRGH). S: Choma, fl. & fr. 14.i.1963, *van Rensburg* 1215 (K; SRGH). **Zimbabwe**. N: Kariba Distr., Bumi Hills Harbour, 15.xii.1964, *D.S. Mitchell* 891 (K; SRGH). W: Umguza Distr., Nyamandhlovu, fl. & fr. 3.ii.1956, *Plowes* 1921 (SRGH). C: Harare, Greendale, 24.ii.1960, *Leach* 9779 (K; SRGH). E: Chimanimani, fl. & fr. 11.vi.1950, *Chase* 2415 (BM; SRGH). S: Mwenezi Distr., Chironga (Chilonga) Irrigation Scheme, fl. & fr. 15.i.1970, *P. Taylor* 10 (K; SRGH). **Malawi**. N: Karonga Distr., 3 km north of Chilumba, fl. & fr. 23.iv.1969, *Pawek* 2300 (K). C: Lilongwe Distr., Agricultural Research Station, fl. & fr. 14.ii.1966, *Salubeni* 411 (K; MAL; SRGH). S: Mangochi Distr., Namwera, fl. & fr. 23.viii.1976, *Pawek* 11654B (K; MAL; MO). **Mozambique**. N: Nampula, fl. & fr. 6.iv.1961, *Balsinhas & Marrime* 349 (BM; K). Z: Lugela Distr., Namagoa Plantations, fl. & fr. no date, *Faulkner Kew* 7 (K). T: Changara Distr., Mudezi (Mudzi) R., fl. & fr. 26.ix.1948, *Wild* 2639 (SRGH). MS: Beira, fl. & fr. ii.1921, *Dummer* 4657 (K). GI: Massinga Distr., 5 km south of Massinga, fl. & fr. ii.1938, *Gomes e Sousa* 2070 (COI; K). M: Inhaca Island, fl. & fr. 29.x.1962, *Mogg* 30034 (K; SRGH).

A very common pantropical weed native to Central America. Cultivated land, roadsides and waste places; 0–1500 m.

2. **Euphorbia indica** Lam., Encycl. Méth., Bot. **2**: 423 (1786). —Boissier in de Candolle, Prodr. **15**, 2: 22 (1862). —Gomes e Sousa, Pl. Menyharth.: 77 (1936). —Andrews, Fl. Pl. Sudan **2**: 71 (1952). —Raju & Rao in Indian J. Bot. **2**: 202 (1979). —S. Carter in F.T.E.A., Euphorbiaceae, part 2: 417, fig. 77 (1988). Type from "East Indies".
   Euphorbia indica var. *angustifolia* Boiss. in de Candolle, Prodr. **15**, 2: 22 (1862). Syntypes from Sudan, Ethiopia and Java.
   Euphorbia indica var. *pubescens* Pax in Bot. Jahrb. Syst. **19**: 117 (1894). Syntypes from Tanzania.
   Euphorbia hypericifolia sensu N.E. Brown in F.T.A. **6**, 1: 498 (1911); in F.C. **5**, 2: 248 (1915). —sensu Robyns, Fl. Sperm. Parc Nat. Alb. **1**: 476 (1948). —sensu Suessenguth & Merxmüller, [Contrib. Fl. Marandellas Distr.] Proc. & Trans. Rhod. Sci. Ass. **43**: 84 (1951). —sensu Hadidi in Bull. Jard. Bot. Belg. **43**: 87 (1973); non L.

Annual herb, spreading or erect, with branches to 50(100) cm long, the whole plant, including the capsule at least when young, sparsely pilose, rarely almost glabrous, often purplish tinged. Leaves with a petiole to 3 mm long; stipules to 1.5 mm long, broadly triangular, laciniate; lamina to 3 × 1.5 cm, ovate, apex rounded, base obliquely rounded, margin obscurely toothed, upper surface often glabrous, rarely both surfaces glabrous. Cymes terminal and axillary, capitate to 1.5 cm in diameter on peduncles to 3 cm long, subtended by a pair of small leaf-like bracts, the whole shortly pilose to glabrous; cyathial peduncles to 2 mm long. Bracts to 2.5 mm long, linear. Cyathia c. 1 × 1 mm with cup-shaped involucres; glands 4, minute, rounded, green, with appendages varying in size to 1 mm in diameter, white; lobes 0.5 mm long, acutely triangular. Male flowers: bracteoles linear; stamens 1.25 mm long. Female flower: ovary pilose or with at least a few short hairs, rarely entirely glabrous; styles 0.5 mm long, suberect, bifid almost to the base. Capsule exserted on a reflexed pedicel 1.5 mm long, 1.5 × 2 mm, acutely 3-lobed. Seeds 1 × 0.75 mm, oblong-conical, with obscure transverse ridges, reddish-brown.

Botswana. N: Mababe Depression, fl. & fr. 14.vi.1978, *P.A. Smith* 2441 (K; SRGH). **Zambia**. N: Mbala Distr., Mpulungu, fl. & fr. 17.iv.1955, *Richards* 5447 (K). C: Mumbwa Distr., Blue Lagoon National Park, 3.7 km southeast of Nyakende House, fl. & fr. 22.ii.1973, *Osborne* 339 (SRGH). S: Gwembe Distr., Lusitu, fl. & fr. 19.v.1960, *Fanshawe* 5686 (K; NDO; SRGH). **Zimbabwe**. N: Mount Darwin Distr., Mukumbura (Mkumbura) R. on road to Tete, fl. & fr. 23.i.1960, *Phipps* 2387 (K; SRGII). C: Harare, National Botanic Garden, 26.iv.1977, *Biegel* 5470 (K; SRGH). E: Chimanimani Distr., Save (Sabi) R., Nyanyadzi, fl. & fr. 3.ii.1948, *Wild* 2499 (K; SRGH). **Malawi**. N: Karonga, fl. & fr. 10.i.1959, *E.A. Robinson* 3136 (K; SRGH). S: Mangochi Distr., Upper Shire Flats, fl. & fr. xii.1893, *Scott-Elliot* 8427 (K). **Mozambique**. T: Tete Distr., "Baroma", Mucanha (N'Kanya) R., Msusa, fl. & fr. 25.vii.1950, *Chase* 2822 (BM; K; SRGH). MS: Chemba Distr., 5 km from Ancueza on Nhacolo (Tambara) road, fl. & fr. 23.iv.1960, *Lemos & Macuácua* 144 (BM; COI; K; SRGH). GI: Chibuto Distr., Maniquenique, fl. & fr. 19.vi.1960, *Lemos & Balsinhas* 145 (COI; K; SRGH).

A common weed introduced originally from India. Floodplain grassland, seasonally waterlogged, usually in black clay soils, or near permanent water; 25–1570 m.

Leaf-size, and especially pubescence, vary according to environmental conditions, plants occurring in drier situations have more hairy, spreading branches and smaller leaves. Glabrescent forms are more common in the Flora Zambesiaca area than in Tanzania and further north, and occasionally some specimens from permanently damp or shaded situations, e.g. the Luangwa Valley, are completely glabrous.

E. hypericifolia, with which *E. indica* has been confused ever since N.E. Brown's account in F.T.A. (1911), is a widespread species of the New World tropics and subtropics. The two species are very closely related, distinguished by the larger usually pubescent capsule of *E. indica* and its stipules which remain separated.

Sect. 2. CHAMAESYCE (S.F. Gray) Rchb., Fl. Germ. Exc.: 755 (1832). —S. Carter in F.T.E.A., Euphorbiaceae, part 2: 418 (1988).

Annual or perennial herbs, erect or prostrate, branching from near the base. Leaves (bracts) opposite, the base obliquely rounded to subcordate; stipules present. Cyathia terminal and pseudoaxillary (one axillary bud develops into a strong branch, reducing the terminal cyathium to an apparently axillary position), solitary or in congested leafy cymes. Involucres bisexual, with 4 glands, rarely unisexual (male) with 5 glands; glands with an entire or lobed petaloid appendage on the outer margin. Stamens just exserted, with subsessile anthers. Perianth of the female flower reduced to a rim below the ovary; styles joined at the base only. Seeds conical, obtusely 4-angled, without a caruncle.

10. Plants erect, glabrous or rarely with a few scattered hairs at the base; stipules to c. 1 mm long, linear and toothed at the broad base · 8. *polycnemoides*
 – Plants prostrate or decumbent, glabrous; stipules c. 1.5 mm long, deeply divided into 3–5 linear teeth · 9. *inaequilatera*
11. Capsules pilose · 11. *delicatissima*
 – Capsules glabrous · 12
12. Stipules linear, to 1.5 mm long, entire; style apices clavate · · · · · · · · · · · · · · 12. *eylesii*
 – Stipules triangular, to 0.5 mm long, toothed; style apices not clavate · · 13. *neopolycnemoides*
13. Capsule subsessile, included within the involucre · · · · · · · · · · · · · · · · · · · 15. *thymifolia*
 – Capsule exserted from the involucre · 14
14. Capsule with a row of hairs along each suture, otherwise glabrous · · · · · · · · 16. *prostrata*
 – Capsule glabrous or sparsely pilose · 15
15. Branches erect, pilose on both surfaces; leaves very distinctly toothed · · · · 10. *serratifolia*
 – Branches prostrate, pilose on the upper surface only; leaves entire or obscurely toothed · 17. *mossambicensis*
16. Gland appendages inconspicuous, minutely crenulate, white · · · · · · · · · · · 18. *schlechteri*
 – Gland appendages conspicuous, entire or shallowly lobed, white, pink or red · · · · · · · 17
17. Leaf apex apiculate; cyathial peduncles 1–25 mm long; gland appendages ± equal in size · 19. *zambesiana*
 – Leaf apex rounded; cyathial peduncles 0.5 mm long; 2 of the gland appendages conspicuously larger · 20. *rubriflora*

3. **Euphorbia tettensis** Klotzsch in Peters, Naturw. Reise Mossambique **6**, part 1: 94 (1861). — Boissier in de Candolle, Prodr. **15**, 2: 49 (1862). —N.E. Brown in F.T.A. **6**, 1: 494 (1911). —Eyles in Trans. Roy. Soc. South Africa **5**: 400 (1916). Type: Mozambique, Tete (Tette), *Peters* s.n. (B†, holotype).

Annual herb, erect to 25 cm tall or spreading and semi-prostrate with branches to 35 cm long, the whole plant including the cyathia and capsules pilose, with short crisped hairs, to densely pilose with long silky white hairs. Leaves subsessile; stipules to 1 mm long, linear; lamina to 18(25) × 10(15) mm, lanceolate with apex acute to broadly ovate with apex obtuse, obliquely cordate at the base, margin entire and often reddened, upper surface often with a large reddish blotch over the midrib. Cymes condensed into very short axillary shoots with much-reduced leaf-like lanceolate bracts 2–3 mm long. Cyathia subsessile, c. 1.5 × 1.5 mm with cup-shaped involucres; glands 4, c. 0.5 mm wide, transversely elliptic, with conspicuous white to deep pink appendages, the 2 glands on either side of the gap through which the capsule is exserted are larger, obliquely extended to 2.5 × 2 mm with an undulate margin; lobes acutely triangular, pilose. Male flowers: bracteoles few, filamentous; stamens 2 mm long. Female flower: ovary pedicellate, densely pilose; styles 1.5 mm long, slender, bifid to about halfway. Capsule exserted on a reflexed pedicel to 1.5 mm long, 1.8 × 2 mm, acutely 3-lobed, truncate at the base, densely pilose. Seeds 1.4 × 1 mm, ovoid with truncate base, 4-angled, with 3–4 shallow but distinct transverse ridges and wrinkles, pale pinkish-brown.

Zambia. C: South Luangwa National Park, near Mfuwe, fl. & fr. 4.iii.1970, *Astle* 5810 (K; NDO; SRGH). E: Petauke Distr., Nyamadzi R., fl. & fr. 25.iii.1955, *Exell, Mendonça & Wild* 1181 (SRGH). S: Gwembe Distr., Maamba, fl. & fr. 9.ii.1978, *Chisumpa* 460 (K; NDO). **Zimbabwe**. N: Hurungwe Distr., 17.5 km south of Chirundu, fl. & fr. 10.v.1956, *McGregor* 24/56 (K; SRGH). W: Umguza Distr., Rochester Farm, fl. & fr. 27.ii.1966, *Leach & Bullock* 13201 (K; SRGH). C: Goromonzi Distr., Domboshawa, fl. & fr. 7.ii.1961, *Rutherford-Smith* 496 (K; SRGH). E: Chipinge Distr., 7 km south of Chisumbanje, fl. & fr. 10.iii.1976, *Pope & Müller* 1523 (K; SRGH). S: Bikita Distr., Save (Sabi) Valley, Elephant Mine, fl. & fr. 12.ii.1966, *Wild* 7541 (K; SRGH). **Malawi**. S: Shire R., Mpatamanga Gorge, fl. & fr. 28.ii.1961, *Richards* 14490 (K; SRGH). **Mozambique**. T: Tete Distr., near Chitima (Estima), fl. & fr. 4.ii.1972, *Macêdo* 4754 (SRGH). MS: Chemba Distr., Ancueza, Chiou, fl. & fr. 20.iv.1960, *Lemos & Macuácua* 126 (COI; K; SRGH).
    As yet only one specimen has been recorded from outside the Flora Zambesiaca area — *Obermeyer* 658, from Zoutpansberg, Northern Province of South Africa. Common herb on sandy stony soils in open deciduous woodland; 50–1550 m.
    Leaf-shape is very variable in this species, influenced apparently by altitude and humidity. Most commonly the leaves are broadly ovate, with a dense pubescence of long silky hairs over the whole plant. Plants from the higher altitudes (600 m and above) tend to have lanceolate leaves and a thinner pubescence of somewhat appressed crisped hairs.

4. **Euphorbia kilwana** N.E. Br. in F.T.A. **6**, 1: 507 (1911). —S. Carter in F.T.E.A., Euphorbiaceae, part 2: 419 (1988). Type from Tanzania.

Euphorbia convolvuloides var. *integrifolia* Pax in Bot. Jahrb. Syst. **43**: 85 (1909). Type as for *E. kilwana*.

Annual herb, semi-prostrate, with branches to 45 cm long, becoming woody towards the base, pilose with short, curved, appressed hairs. Leaves with a petiole to 1.5 mm long; stipules to 1 mm long, linear, pilose; lamina to 25 × 8 mm, ovate-oblong, apex obtuse, margin entire, glabrous or almost so above, thinly pilose beneath. Cymes clustered into short leafy axillary shoots; cyathia c. 1.3 × 1.25 mm with pilose barrel-shaped involucres; peduncles to 1 mm long; glands 4, minute, circular, with minute red appendages; lobes minute, triangular. Male flowers: bracteoles filamentous; stamens 1.5 mm long. Female flower: ovary pedicellate, pilose; styles 0.2–0.3 mm long, erect, bifid almost to the base. Capsule exserted on a reflexed pedicel to 1.5 mm long, 1.8 × 2 mm, acutely 3-lobed with a truncate base, densely pilose with short, curved appressed hairs. Seeds 1.1 × 0.8 mm, ovoid, 4-angled, with shallow transverse ridges, pinkish-brown.

Zimbabwe. E: Chipinge Distr., east bank of lower Save (Sabi) R., 1.5 km northeast of Hippo Mine, fl. & fr. 12.iii.1957, *Phipps* 582 (SRGH). **Malawi**. S: Chikwawa Distr., banks of Shire R. above Elephant Marsh, fl. & fr. ii.1887, *L. Scott* s.n. (K). **Mozambique**. GI: valley of Limpopo R., near Xai-Xai, 21.i.1981, *Jansen* 7674 (K).
Also in eastern Tanzania. River bank alluvium, usually seasonally flooded, in clay soils; c. 10–500 m.

5. **Euphorbia lupatensis** N.E. Br. in F.T.A. **6**, 1: 514 (1911). —S. Carter in F.T.E.A., Euphorbiaceae, part 2: 420 (1988). Type: Mozambique, Lupata, fl. & fr. 20.iv.1860, *Kirk* s.n. (K, holotype).

Erect annual herb to 30 cm high; branches spreading, pubescent with short, curved appressed hairs. Leaves with a petiole to 2 mm long; stipules 1–2 mm long, linear, entire; lamina to 30(45) × 5(7) mm, linear to linear-lanceolate, acute to obtuse at the apex, obliquely rounded at the base, margin entire or with a few obscure teeth, upper surface glabrous when mature, lower surface with scattered appressed hairs. Cymes condensed into very short axillary shoots with much-reduced leaf-like pubescent lanceolate bracts 2–3 mm long; cyathia subsessile, c. 1.5 × 1.5 mm with cup-shaped pubescent involucres; glands 4, to 0.5 mm wide, transversely elliptic, with usually conspicuous white or pink appendages, the 2 glands on either side of the gap through which the capsule is exserted are often larger, obliquely extended to 2 × 1.5 mm with undulate margin; lobes acutely triangular, ciliate. Male flowers: bracteoles deeply laciniate; stamens 2 mm long. Female flower: ovary pedicellate, densely pubescent; styles 0.5–1.5 mm long, slender, bifid for 0.5 mm at the apex. Capsule exserted on a reflexed pedicel c. 2 mm long, 2.2 × 2.4 mm, obtusely 3-lobed, broader at the base, densely pubescent with short curved appressed hairs. Seeds (1.2)1.7 × 1 mm, oblong-conical, 4-angled, with numerous distinct transverse ridges and wrinkles, pinkish-grey.

Zambia. C: South Luangwa National Park, fl. & fr. 14.ii.1967, *Prince* 169 (K; NDO). E: Petauke Distr., 4 km west of Kachalola on Great East Road., fl. & fr. 17.iii.1959, *Robson* 1739 (K; SRGH). S: Gwembe Distr., Maamba, fl. & fr. 9.ii.1978, *Chisumpa* 451 (K). **Zimbabwe**. N: Lake Kariba, Ruaru Island, fl. & fr. 9.i.1963, *Whellan* 2004 (MAL; SRGH). W: Hwange National Park, Shapi Camp, fl. & fr. 27.ii.1967, *Rushworth* 271 (SRGH). E: Chipinge Distr., east Save (Sabi) between Musasvi R. and Cilariati (Cikariati) R., fl. 22.i.1957, *Phipps* 109 (K; SRGH). **Malawi**. S: Mwanza Distr., Neno, Blantyre–Mwanza road, fl. & fr. 9.ii.1938, *Lawrence* 553 (K). **Mozambique**. MS: Cahora Bassa Distr., 21 km from Chicoa towards Chetima, fl. & fr. 30.vi.1949, *Barbosa & Carvalho* 3404 (K).
Also in southeast Tanzania. Wooded grassland and open woodland, on sandy clay soils; 300–1025 m.
Since the type appears to be a depauperate specimen, this species is unsatisfactorily defined and appears to be extremely variable. Leaves on the type specimen are to 35 × 4 mm, linear, with entire margins, but all intermediates occur between this and plants with leaves usually about 20 × 4 mm, linear-lanceolate, and with a few obscure teeth. Specimens have been collected from Zambia, South Luangwa National Park, with leaves to 22 × 7 mm, ovate, and with obviously toothed margins (*Astle* 4369, *Prince* 57 and 72). The size of the glandular appendages can vary on one individual, but those on collections from the east of the distribution pattern

(Mozambique, Malawi, Zambia E) are generally larger than on those from further west. In addition, capsules and seeds of the eastern plants are smaller than those of plants from Zimbabwe and Zambia C and S.

A specimen from Mozambique, 10 km from Mutuali on the road to Mulema, *Gomes e Sousa* 4073, has leaves to 20 × 7 mm, ovate, distinctly toothed, with small capsules c. 1.7 × 1.8 mm and small seeds 1.2 × 0.9 mm. A poor specimen from the same region, 21 km west of Ribáuè, *Leach & Rutherford-Smith* 10991 is similar but has less distinctly toothed leaves.

An isolated collection from Botswana SE, 60 km northwest of Serowe, *Wild & Drummond* 7281, has leaves to 20 × 6 mm, ovate-lanceolate, distinctly toothed, and with conspicuous stipules 2 mm long, linear. Another, from Botswana N, 69 km west of Nokaneng, *Wild & Drummond* 6891, is entirely covered with long silky hairs, including the capsule. Its branches are no more than 15 cm long, the leaves to 20 × 3.5 mm, lanceolate, are toothed at the apices and the glandular appendages are very small. Both these specimens, especially the latter, may eventually prove to be distinct at specific level.

Rather than attempting to define a number of distinct taxa within this complex, it seems best to regard it as a heterogeneous species, at least until considerably more material is available for comparison, especially from the area of the type collection.

6. **Euphorbia karibensis** S. Carter in Kew Bull. **45**: 330 (1990). TAB. **72**, fig. A. Type: Zimbabwe, Kariba Distr., summit of Namsowa (Nyasau), fl. & fr. 15.iii.1959, *Chase* 7066 (K, holotype; BM; SRGH).

Erect annual herb to 30 cm high with a stem to 3 mm thick becoming woody; branches spreading, sparsely pubescent to almost glabrous, with short, curled hairs. Leaves glabrous with a petiole 0.5–1 mm long; stipules to 1.5 mm long, linear, entire; lamina to 45 × 3 mm, linear or to 30 × 7 mm and lanceolate, acute at the apex, obliquely rounded at the base, entire on the margins. Cyathia terminal and pseudoaxillary, congested on short leafy shoots, solitary on peduncles 0.5–1 mm long, c. 1.3 × 1.3 mm with cup-shaped involucres; glands 4, to 0.5 mm wide, transversely elliptic, with white to pink appendages, the 2 on either side of the gap through which the capsule is exserted, slightly larger to 0.8 × 1 mm; lobes acutely triangular, ciliate. Male flowers: bracteoles deeply laciniate; stamens 2.4 mm long. Female flower: ovary pedicellate, glabrous; styles 1.3 mm long, joined at the base, scarcely bifid at the apex. Capsule exserted on a reflexed pedicel to 2.5 mm long, 2 × 2.3 mm, 3-lobed with truncate base. Seeds 1.5 × 1 mm, oblong-conical, 4-angled, with 3 or 4 pronounced transverse ridges and grooves, pale pinkish-brown to pale grey.

Zimbabwe. N: Binga Distr., Kariba, Sinamwenda Zone, fl. & fr. 8.iv.1966, *Jarman* 18 (K; SRGH).

So far collected only from the hills along Lake Kariba shore and above the Kariba Dam. Stony soil on hillsides; 760–1100 m.

A further specimen, *Kornaś* 1599 from Zambia, Mkushi Distr., Muchinga Escarpment, Lunsemfwa Wonder Gorge, is included here with reservation. The duplicate deposited in Kew is poor, but appears to be etiolated, differing from other specimens by elongated gland appendages, to 2 × 1 mm, and by styles 2.5 mm long with undivided apices.

7. **Euphorbia spissiflora** S. Carter in Kew Bull. **45**: 331 (1990). TAB. **72**, fig. B. Type: Zimbabwe, Nhongo, 8 km north of Gokwe, fl. & fr. 6.iii.1964, *Bingham* 1158 (K, holotype; SRGH).

Erect annual herb to 50 cm high; branches spreading, glabrous or occasionally minutely pubescent, becoming woody at the base, longitudinally ridged towards the apex. Leaves with a petiole 1–2 mm long; stipules to 1.5 mm long, linear from a broad base, entire; lamina glabrous, to 40 × 5 mm, linear or linear-lanceolate, obtuse at the apex, obliquely rounded at the base, denticulate on the margins especially towards the apex, rarely entire. Cyathia terminal and pseudoaxillary, congested on short leafy shoots, solitary on peduncles 0.5–1.5 mm long, c. 1.3 × 1.3 mm with cup-shaped involucres; glands 4, to 0.5 mm wide, transversely elliptic, with pink or white appendages variable in size to 0.6 × 1 mm; lobes minute, acutely triangular. Male flowers: bracteoles filamentous; stamens 1.5 mm long. Female flower: ovary pedicellate, glabrous; styles 0.5 mm long, joined at the base, bifid almost to the base. Capsule exserted on a pedicel c. 1.5 mm long, 1.5 × 1.8 mm, acutely 3-lobed. Seeds 1.1 × 0.7 mm, oblong-conical, 4-angled with shallow transverse wrinkles, pale pinkish-grey.

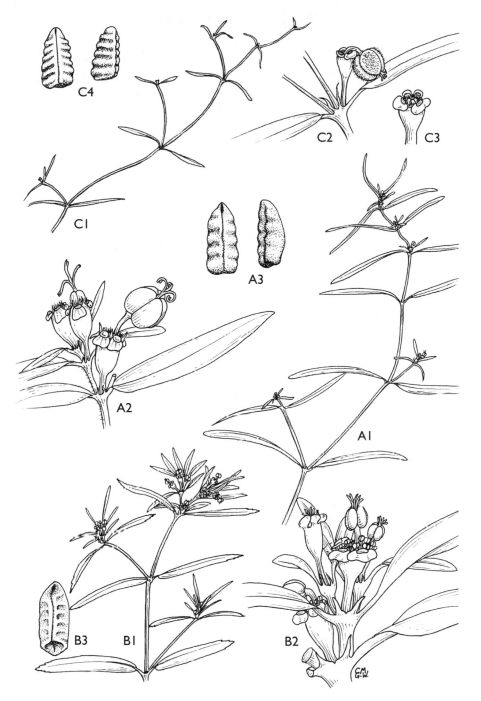

Tab. 72. A. —EUPHORBIA KARIBENSIS. A1, fertile branch (× ²/₃), from *Chase* 7066; A2, inflorescence (× 6); A3, seeds (× 12), A2 & A3 from *Jarman* 18. B. —EUPHORBIA SPISSIFLORA. B1, fertile branch (× ²/₃); B2, inflorescence (× 6); B3, seed (× 12), B1–B3 from *Bingham* 1158. C. —EUPHORBIA DELICATISSIMA. C1, fertile branch (× ²/₃); C2, leaf-bases and fruiting cyathium (× 6); C3, cyathium showing petaloid gland appendages (× 6); C4, seeds (× 12), C1–C4 from *Grosvenor* 478. Drawn by Christine Grey-Wilson. From Kew Bull.

Botswana. N: Central Distr., Pandamatenga, 11.iv.1991, *M. Phillips* 42 (K). **Zambia**. B: Sesheke Distr., Machili, fl. & fr. 10.ii.1961, *Fanshawe* 6232 (NDO; SRGH). S: Mumbwa Distr., fl. 1911, *Macaulay* 358 (K). **Zimbabwe**. N: Nhongo, 8 km north of Gokwe on road to Copper Queen, fl. & fr. 6.iii.1964, *Bingham* 1158 (K; SRGH). W: Hwange Distr., Matetsi, fl. & fr. 6.iv.1978, *Gonde* 143 (SRGH). C: Chegutu Distr., Poole Farm, fl. & fr. 21.iii.1948, *R.M. Hornby* 2908 (K; SRGH).

Known only from southern Zambia and northwest Zimbabwe. In grassland on stony clay soils; c. 850–1200 m.

8. **Euphorbia polycnemoides** Boiss. in de Candolle, Prodr. **15**, 2: 46 (1862). —N.E. Brown in F.T.A. **6**, 1: 506 (1911). —Andrews, Fl. Pl. Sudan **2**: 71 (1952). —Keay in F.W.T.A., ed. 2, **1**: 421 (1958). —S. Carter in F.T.E.A., Euphorbiaceae, part 2: 428, fig. 79/5–7 (1988). Syntypes from Sudan and Ethiopia.

Annual herb with branching stems erect to 35 cm high, glabrous except occasionally for a few long scattered hairs on the basal branches and leaves, the whole plant often tinged red. Leaves subsessile; stipules c. 1 mm long, linear and toothed at the broad base; lamina to 15×4 mm, lanceolate, rounded and apiculate at the apex, obliquely rounded to subcordate at the base, margin entire in the lower half, toothed in the upper half, obscurely so on smaller leaves. Cyathia terminal and pseudoaxillary on short leafy shoots, solitary on peduncles 1 mm long, 1×1 mm with cup-shaped involucres; glands 4, minute, transversely elliptic, often red, with minute pink or red lobed appendages; involucral lobes 0.5 mm long, triangular, toothed. Male flowers: few, bracteoles filamentous; stamens 1 mm long. Female flower: ovary pedicellate; styles 0.5 mm long, bifid almost to the base. Capsule exserted on a reflexed pedicel 1.5 mm long, 1.5×1.5 mm, acutely 3-lobed. Seeds 1 $\times 0.5$ mm, oblong-conical, with 3–4 distinct transverse wrinkled ridges and grooves, pinkish-brown.

Zambia. S: Livingstone Distr., fl. & fr., no date, *Macaulay* 562 (K). **Malawi**. N: Karonga Distr., lake shore, 8 km north of Livingstonia junction, fl. & fr. 24.ii.1978, *Pawek* 13880 (K; MAL; MO). C: Salima Distr., Chitala, Lilongwe road, fl. & fr. 12.ii.1959, *Robson* 1556 (K; SRGH).

Also in Ethiopia and Sudan westwards to Nigeria, and south through Dem. Rep. Congo and southern central Tanzania. *Brachystegia* woodland; 475–1000 m.

9. **Euphorbia inaequilatera** Sond. in Linnaea **23**: 105 (1850). —N.E. Brown in F.C. **5**, 2: 246 (1915). —Bremekamp & Obermeyer [Scientific Results of the Vernay-Lang Kalahari Expedition, March to September, 1930], in Ann. Transvaal Mus. **16**: 421 (1935). —Robyns, Fl. Sperm. Parc Nat. Alb. **1**: 478 (1948). —Andrews, Fl. Pl. Sudan **2**: 72 (1952). —Agnew, Upland Kenya Wild Fl.: 221 (1974). —S. Carter in F.T.E.A., Euphorbiaceae, part 2: 426 (1988). Type from South Africa.

Annual herb, rarely perennial, prostrate and much-branched up to 50 cm in diameter or sometimes decumbent with branches up to 30 cm long; branches longitudinally ridged especially when dry, completely glabrous. Leaves with a petiole to 1.5 mm long; stipules to 1.5 mm long, deeply divided into 3–5 linear points; lamina to 10×6 mm, ovate to occasionally lanceolate, obtuse at the apex, very obliquely rounded to subcordate at the base, margin minutely toothed, often very obscurely so. Cyathia terminal and pseudoaxillary on short leafy shoots, solitary on peduncles 1 mm long, 1×1 mm with cup-shaped involucres; glands 4, minute, transversely elliptic, red with small lobed pink or white appendages; involucral lobes minute, triangular, sharply toothed. Male flowers: bracteoles laciniate; stamens 1.25 mm long. Female flower: ovary pedicellate; styles 0.5 mm long, spreading, bifid almost to the base. Capsule exserted on a reflexed pedicel 2 mm long, 1.5×1.75 mm, acutely 3-lobed with the angles often purple-tinged. Seeds 1.25×0.75 mm, oblong-conical, with shallow transverse wrinkles and pits, greyish-brown.

Var. **inaequilatera**
> *Euphorbia sanguinea* Boiss. in de Candolle, Prodr. **15**, 2: 35 (1862). —N.E. Brown in F.T.A. **6**, 1: 508 (1911). Syntypes from N Yemen and Ethiopia.
> *Euphorbia sanguinea* var. *natalensis* Boiss. in de Candolle, Prodr. **15**, 2: 35 (1862) nom. illegit. Type as for *E. inaequilatera*.

Euphorbia sanguinea var. *intermedia* Boiss. in de Candolle, Prodr. **15**, 2: 35 (1862). —N.E. Brown in F.T.A. **6**, 1: 508 (1911). Syntypes from Ethiopia.
Euphorbia inaequalis N.E. Br. in F.T.A. **6**, 1: 512 (1911). Type from Somalia.

Annual herb, with fibrous roots.

Botswana. N: Dindinga Island, fl. & fr. 27.iii.1975, *P.A. Smith* 1316 (K; SRGH). SW: 4 km north of Dondong Borehole, fl. & fr. 1.ii.1977, *Skarpe* S-127 (K; SRGH). SE: Tlalamabele–Mosu area, near Sua (Soa) Pan, fl. & fr. 15.i.1974, *Ngoni* 341 (K; SRGH). **Zambia**. B: Masese, fl. & fr. 14.i.1961, *Fanshawe* 6134 (NDO). C: Iolanda, north bank of Kafue R., near Kafue town, fl. & fr. 14.iii.1965, *E.A. Robinson* 6426 (K; SRGH). S: Mazabuka Distr., Kalomo, fl. & fr. 1.i.1958, *E.A. Robinson* 2555 (K; SRGH). **Zimbabwe**. N: Zvimba Distr., 1.6 km north of Mutorashanga, fl. & fr. 1.i.1975, *Wild* 7984 (K; SRGH). W: Matopos Distr., Lucydale Farm, fl. & fr. 15.xii.1968, *Leach & Cannell* 14158 (K; SRGH). C: Harare, Greendale, fl. & fr. 23.ii.1960, *Leach* 9778 (K; SRGH). **Malawi**. C: Dedza Distr., Chongoni Forest Reserve, fl. & fr. 19.iii.1971, *Salubeni* 1527 (K; MAL; SRGH).

A common weed, from the Arabian peninsula through Somalia, Ethiopia, East Africa and southwestwards to Angola and South Africa, but excluding southeastern Tanzania and areas east of Lake Malawi. Open patches amongst grass on usually seasonally wet soils; 600–1460 m.

The degree to which the leaves are toothed is very variable, and becomes less pronounced towards the south of the species distribution. In Botswana especially, where it appears to be particularly common, leaves are generally smaller than on plants to the north in Zambia and East Africa, and are often entire except for slight indentations of the cartilaginous margin at the apex of at least a few leaves.

One specimen from Zambia, Namwala Distr., Lochinvar National Park, NNE of Chunga, *van Laviern, Sayer & Rees* 303, matches other material of *E. inaequilatera* except for long, very scattered hairs on leaves and capsules and smooth brown seeds. It may represent a distinct species but more gatherings are needed before this can be decided.

Var. **perennis** N.E. Br. in F.C. **5**, 2: 246 (1915). Types from South Africa.

An apparently perennial herb, probably short-lived, with a woody root, c. 5 mm thick; branches prostrate to 15 cm long. Leaves to 6 × 4 mm, ovate, margin entire or finely toothed. Otherwise as for var. *inaequilatera*.

Botswana. SE: between Gaborone and Francistown, fl. & fr. ix.1967, *Lambrecht* 314 (K); Digkatlong Ranch, fl. & fr. 5.ii.1977, *O.J. Hansen* 3028 (K).

These are the only specimens I have seen from the Flora Zambesiaca area, this variety apparently being otherwise confined to nearby regions of the Cape Province, Northern and North-West Provinces of South Africa. In sandy soils; c. 1030 m.

One of the syntypes of this variety, *Burke* 507 from the Sand R., consists of a collection of several small plants. Three of these have leaves with an entire margin, the rest have toothed leaves, a feature which is variable throughout the range of the species as a whole. Of the 2 specimens from Botswana, *Lambrecht* 314 has toothed leaves, and *Hansen* 3028 has leaves with the margin entire.

10. **Euphorbia serratifolia** S. Carter in Kew Bull. **35**: 413 (1980); in F.T.E.A., Euphorbiaceae, part 2: 425 (1988). Type from Tanzania.

Annual herb to 25 cm high with spreading branches, the whole plant sparsely covered with long spreading hairs, the upper side of the branches with shorter appressed hairs in addition. Leaves with a petiole to 2 mm long; stipules 0.75 mm long, deeply divided into 2–3 linear teeth; lamina to 18 × 12 mm, ovate, subacute at the apex, very obliquely subcordate at the base, distinctly toothed with teeth apparently gland-tipped on the margin. Cyathia terminal and pseudoaxillary on leafy shoots, solitary on peduncles to 2.5 mm long, 1 × 1 mm with cup-shaped involucres; glands 4, c. 0.3 mm wide, transversely elliptic, red, with obvious, deeply 3–6-lobed pink or red appendages; involucral lobes broadly triangular, margin ciliate. Male flowers; bracteoles linear, deeply divided; stamens 1 mm long. Female flower: ovary pedicellate; styles 0.3 mm long, spreading, bifid almost to the base. Capsule exserted on a reflexed pedicel 1.25 mm long, 1.5 × 2 mm, 3-lobed with broadly truncate base. Seeds 1 × 0.75 mm, ovoid, with 3 distinct transverse ridges, reddish-brown.

Malawi. N: 32 km northwest of Rumphi on M1, fl. & fr. 11.iii.1978, *Pawek* 14043 (K). C: Chitala Escarpment, fl. & fr. 14.ii.1959, *Robson* 1591 (BM; K; MAL; SRGH).

Also in southern Tanzania. Sandy stony soil in open *Brachystegia* woodland; 600–1400 m.

11. **Euphorbia delicatissima** S. Carter in Kew Bull. **45**: 328 (1990). TAB. **72**, fig. C. Type: Zimbabwe, Harare Distr., Christon Bank, Botanic Garden Extension, fl. & fr. 12.iii.1969, *Grosvenor* 478 (K, holotype; SRGH).
  Euphorbia sp. (*Eyles* 314) of Eyles in Trans. Roy. Soc. South Africa **5**: 400 (1916).

Glabrous annual herb, erect to 30 cm high; branches lax, spreading, very slender. Leaves with a petiole 0.5 mm long; stipules 0.5–0.75 mm long, filiform; lamina to 30 × 2 mm, linear, apiculate at the apex, obliquely rounded at the base and entire on the margin. Cyathia terminal and pseudoaxillary, solitary on peduncles 0.5–1.5 mm long, 1.25 × 1.25 mm with cup-shaped involucres; glands 4, to 0.5 mm wide, transversely elliptic, with white to deep pink appendages, the 2 on either side of the gap through which the capsule is exserted larger, obtusely extended to 1.5 × 0.8 mm, usually smaller; lobes minute, acutely triangular. Male flowers: bracteoles few, filiform; stamens 1.5 mm long. Female flower: ovary pedicellate, pilose; styles 0.75 mm long, joined at the base, spreading, bifid for two-thirds, with slightly thickened apices. Capsule exserted on a reflexed pedicel to 1.8 mm long, 1.6 × 1.6 mm, acutely 3-lobed with truncate base, pilose except around the apex. Seeds 1.4 × 0.8 mm, oblong-conical, 4-angled, with numerous fine transverse ridges and grooves, pale pinkish-brown to pale grey.

Zimbabwe. N: Zvimba Distr., west side of Mvurwi (Umvukwe) Range, 24 km south of Kildonan, fl. & fr. 25.ii.1959, *Drummond, Jackson & Phipps* 5847 (EA; K; SRGH); Mazowe Distr., Christon Bank, Mazowe (Mazoe) R. headwaters, fl. & fr. 22.v.1966, *Loveridge* 1512 (K; SRGH).
  Recorded only from the vicinity of the Great Dyke and hills about the Mazowe R. headwaters. On stony hillsides amongst grass; 1400–1700 m.
  Easily distinguished from *E. eylesii* by the pilose capsule.

12. **Euphorbia eylesii** Rendle in J. Bot. **43**: 52 (1905). —N.E. Brown in F.T.A. **6**, 1: 512 (1911) pro parte excl. *Monro* 910. —Eyles in Trans. Roy. Soc. South Africa **5**: 399 (1916) pro parte excl. *Monro* 910. —Wild in Clark, Victoria Falls Handb.: 148 (1952). —S. Carter in Kew Bull. **45**: 327 (1990). TAB. **73**, fig. A. Type: Zimbabwe, Deka Siding on Bulawayo–Victoria Falls railway line, fl. v.1904, *Eyles* 130 (BM, holotype; SRGH).
  Euphorbia leshumensis N.E. Br. in F.T.A. **6**, 1: 513 (1911). Syntypes: Botswana, Lyshuma (Leshumo) Forest, fl. & fr. no date, *Holub* s.n. (K); Zambia, Sesheke, fl. & fr. 1910, *Macaulay* 423 (K).
  Euphorbia neopolycnemoides sensu Merxmüller, Prodr. Fl. SW. Afrika, fam. part 67: 31 (1967), non Pax & K. Hoffm.

Glabrous annual herb, erect to 50 cm high, laxly branched. Leaves with a petiole 1 mm long; stipules to 1.5 mm long, linear; lamina to 30(50) × 3(3.5) mm, linear, apiculate at the apex, obliquely rounded at the base, entire on the margin. Cyathia terminal and pseudoaxillary, solitary on peduncles to 2 mm long, c. 2 × 2 mm, with cup-shaped involucres; glands 4, c. 0.5 mm wide, transversely elliptic, with conspicuous white to pink appendages, the 2 on either side of the gap through which the capsule is exserted very much larger, obliquely extended to 4 × 4 mm; lobes acutely triangular, ciliate. Male flowers: bracteoles laciniate; stamens 2.5 mm long. Female flower: ovary pedicellate; styles c. 1.5 mm long, joined at the base, bifid to halfway with much thickened clavate apices. Capsule exserted on a reflexed pedicel to 3 mm long, 1.8 × 2 mm, deeply 3-lobed. Seeds 1.5 × 0.8 mm, oblong-conical, with 3 or 4 pronounced transverse ridges, pinkish-grey.

Caprivi Strip. 29 km west of Katima Mulilo, fl. & fr. 12.ii.1969, *de Winter* 9130 (K). **Botswana**. N: Pandamatenga, road to Bushman Pits, fl. & fr. 27.iii.1961, *Richards* 14878 (K; SRGH). **Zambia**. B: Senanga Distr., Sioma, fl. & fr. 1.ii.1975, *Brummitt, Chisumpa & Polhill* 14200 (K). C: 1.5 km north of Chirundu, *Rutherford-Smith* 636 (SRGH). S: Livingstone, 3.ii.1961, *Fanshawe* 6173 (K; NDO; SRGH). **Zimbabwe**. N: Gokwe Distr., at Charama road turn-off, fl. & fr. 25.iv.1962, *Bingham* 237 (K). W: Hwange Distr., Matetsi, fl. & fr. iii.1918, *Eyles* 1282 (BM; K; SRGH).
  Known only from these regions apart from the northeastern corner of Namibia. In deciduous woodland on Kalahari Sand; 300–1250 m.
  The size of the petaloid appendages on the involucral glands varies considerably, even on one plant, as does the size of the leaves. This species is easily distinguished from the similar *Euphorbia neopolycnemoides* by its entire stipules and distinctly clavate stigmas.

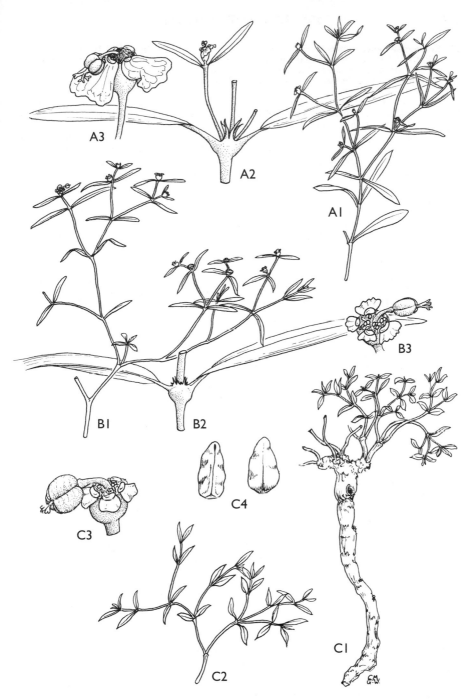

Tab. 73. A. —EUPHORBIA EYLESII. A1, fertile branch (× ⅔); A2, part of branch showing stipules (× 2); A3, cyathium showing gland appendages (× 4), A1–A3 from *de Winter* 9130. B. —EUPHORBIA NEOPOLYCNEMOIDES. B1, fertile branch (× ⅔); B2, part of branch showing stipules (× 2); B3, cyathium (× 4), B1–B3 from *Leach* 12024. C. — EUPHORBIA ZAMBESIANA var. ZAMBESIANA. C1, habit (× ⅔), from *Milne-Redhead* 829; C2, leafy branch (× ⅔); C3, cyathium (× 4), C2 & C3 from *Brenan & Greenway* 5957; C4, seeds (× 9), from *Milne-Redhead* 829. Drawn by Christine Grey-Wilson.

13. **Euphorbia neopolycnemoides** Pax & K. Hoffm. in Bot. Jahrb. Syst. **45**: 240 (1910). —N.E.
Brown in F.C. **5**, 2: 249 (1915). TAB. **73**, fig. B. Type from South Africa (Transvaal).
    Euphorbia arabica var. *latiappendiculata* Pax in Bot. Jahrb. Syst. **43**: 85 (1909). Type from
South Africa (Transvaal).
    Euphorbia eylesii sensu Eyles in Trans. Roy. Soc. South Africa **5**: 399 (1916), quoad *Monro* 910.

Glabrous annual herb, erect to 25 cm tall with laxly spreading branches. Leaves
with a petiole c. 1 mm long; stipules 0.5 mm long, divided from a broad base into
3–5 linear teeth; lamina to 25(30) × 3.5(5) mm, linear to linear-lanceolate, apex
apiculate, base obliquely rounded, margin entire. Cyathia terminal and
pseudoaxillary, solitary on peduncles to 2 mm long, c. 1.5 × 1.5 mm with cup-shaped
involucres; glands 4, 0.5 mm wide, transversely elliptic, with fairly conspicuous white
to pink appendages, the 2 on either side of the gap through which the capsule is
exserted are slightly larger and up to 1.5 × 2 mm; lobes triangular, ciliate. Male
flowers: bracteoles laciniate; stamens 1.5 mm long. Female flower: ovary
pedicellate; styles 1.25 mm long, joined at the base, bifid for one-third. Capsule
exserted on a recurved pedicel to 3.5 mm long, 2.5 × 2.5 mm, acutely 3-lobed with
truncate base. Seeds 1.3 × 0.9 mm, oblong-conical with c. 3 transverse ridges and
grooves, reddish-brown.

    Botswana. N: near Tshesebe (Tsessebe), fl. & fr. 8.iii.1965, *Wild & Drummond* 6812 (K;
SRGH). SE: Kweneng Distr., Gabane Hills between Gabane and Kumakwane, 24°39'S, 25°47'E,
fl. & fr. 5.xi.1978, *O.J. Hansen* 3531 (K; SRGH). **Zimbabwe**. W: Matobo Distr., Besna Kobila
Farm, fl. & fr. i.1959, *Miller* 5729 (K; SRGH). E: Chipinge Distr., fl. 19.i.1957, *Phipps* 30 (K;
SRGH). S: Gwanda Distr., 45 km from Koodoovale Motel on road to Thuli (Tuli) Breeding
Station, fl. & fr. 20.iii.1959, *Drummond* 5875 (K; SRGH); Beitbridge Distr., Nulli Range on
Tshiturapadsi (Chiturupadzi) road, fl. & fr. 10.i.1961, *Leach* 10670 (K; SRGH). **Mozambique**.
M: Lebombo Mts., fl. & fr. 8.i.1929, *Hutchinson* 2547 (K).
    Limited in distribution to the eastern parts of Botswana and southern parts of Zimbabwe
extending east into Mozambique and south into the Transvaal and KwaZulu-Natal. On stony
ground amongst grass in mopane woodland; 300–1460 m.
    The size of the petaloid appendages on the involucral glands varies considerably even on one
plant, but the 2 larger ones are never as big as those found in *Euphorbia eylesii*, a species with a
very similar appearance. *Euphorbia neopolycnemoides* can be easily distinguished by its shorter
stipules divided into several linear teeth, and by its shorter styles without the clavate apices
typical of *Euphorbia eylesii*.

14. **Euphorbia serpens** Kunth in Humboldt, Bonpland and Kunth, Nov. Gen. Sp. **2**: 52 (1817).
—N.E. Brown in F.T.A. **6**, 1: 511 (1911). —S. Carter in F.T.E.A., Euphorbiaceae, part 2: 421
(1988). Type from Venezuela.
    Euphorbia minutiflora N.E. Br. in F.T.A. **6**, 1: 1036 (1913). Type: Zimbabwe, Victoria Falls,
vii.1908, *Schwarz* in *Herb. Bolus* 13027 (K, holotype).

Much-branched prostrate annual herb, with branches to 10(20) cm long,
completely glabrous. Leaves with a petiole 0.2–0.5 mm long; stipules united at the
swollen nodes, 0.5 × 1 mm, triangular with toothed margins; lamina 1–3 mm long,
subcircular, obliquely cordate at the base, margins entire. Cyathia terminal and
pseudoaxillary on short leafy shoots, solitary on peduncles to 2 mm long, 0.7 × 0.7
mm with cup-shaped involucres; glands 4, minute, transversely elliptic, with small
white shallowly lobed appendages; involucral lobes minute, fringed. Male flowers:
bracteoles laciniate; stamens 0.7 mm long. Female flower: perianth evident as a 3-
lobed rim below the shortly pedicellate ovary; styles 0.2 mm long, spreading, bifid
almost to the base. Capsule exserted on a reflexed pedicel to 1.8 mm long, 1.5 × 1.8
mm, obtusely 3-lobed with truncate base, yellowish-green. Seeds 1 × 0.6 mm, oblong-
conical, smooth, greyish-pink.

    Zimbabwe. N: Kariba Distr., west end of Kariba Gorge, fl. & fr. 25.xi.1953, *Wild* 4263 (K;
SRGH). W: Hwange Distr., Deka R., fl. & fr. 21.vi.1934, *Eyles* 7964 (BM; K; SRGH). C: Harare,
Industrial Site, fl. & fr. 23.iv.1976, *Pope & Biegel* 1539 (K; SRGH). S: Chiredzi Distr., Hippo
Valley Estate, Section 17, fl. & fr. 27.xii.1971, *P. Taylor* 201 (K; SRGH). **Malawi**. S: Nsanje Distr.,
Makanga Experimental Station, fl. & fr. 19.iii.1960, *Phipps* 2552 (K; SRGH). **Mozambique**. MS:
Gorongosa Distr., Parque Nacional da Gorongosa (Gorongosa Nat. Park), Urema flood plains,
fl. & fr. vii.1970, *Tinley* 1940 (SRGH). GI: Chicualacuala Distr., Dumela, fl. & fr. 30.iv.1961,
Drummond & Rutherford-Smith 7612 (K; SRGH). M: Namaacha Mts., fl. & fr. 20.xi.1966, *Moura*
117a (COI).

A pantropical weed, but not common in Africa, introduced originally most probably into W Africa. Disturbed and waste ground, often near water-courses; 0–1450 m.

15. **Euphorbia thymifolia** L., Sp. Pl.: 454 (1753). —Boissier in de Candolle, Prodr. **15**, 2: 47 (1862). —Keay in F.W.T.A., ed. 2, **1**: 421 (1958). —S. Carter in F.T.E.A., Euphorbiaceae, part 2: 420 (1988). Type from India.
    Euphorbia afzelii N.E. Br. in F.T.A. **6**, 1: 506 (1911). Type from Sierra Leone.

Prostrate densely branching annual herb, whole plant tinged reddish-brown; branches to 25 cm long, glabrous on the underside, densely pilose above with curved appressed hairs. Leaves with a petiole 0.5 mm long; stipules to 1.25 mm long, linear, often deeply 2–3-toothed, pilose; lamina to 8 × 4 mm, ovate, obtuse at the apex, obliquely subcordate at the base, shallowly toothed at the margin, upper surface glabrous, lower surface with long scattered hairs.   Cyathia terminal and pseudoaxillary on congested leafy shoots, solitary, subsessile, c. 0.5 × 0.5 mm, with funnel-shaped involucres, pilose; glands 4, minute, subcircular, red, with often almost invisible red appendages; lobes minute, triangular, ciliate. Male flowers very few (5 or less): bracteoles reduced to 1 or 2 threads; stamens 0.8 mm long. Female flower: ovary subsessile; styles 0.6 mm long, erect, bifid to halfway.   Capsule subsessile, splitting the involucre during development, 1 × 1 mm, 3-lobed with a truncate base, pilose with short appressed hairs. Seeds 0.6 × 4 mm, conical, sharply 4-angled, with shallow transverse ridges and grooves, reddish-brown.

Zambia. W: Kitwe, fl. & fr. 13.vi.1963, *Mutimushi* 314 (K; NDO; SRGH). E: Chipata Distr., Luangwa R. east bank, 32 km south of Mfuwe, fl. & fr. 2.vi.1970, *Abel* 102 (SRGH). **Mozambique**. Z: Lugela Distr., Namagoa, fl. & fr. 6.v.1948, *Faulkner* 259 partly (K, mixed in with *E. prostrata*).
A pantropical weed introduced into West Africa and found eastwards to Central African Republic and Dem. Rep. Congo; introduced apparently recently into Tanzania and Zambia. Disturbed ground in sandy soil; 500–1300 m. [The plant mixed in with a gathering of *E. prostrata*, *Faulkner* 259 from Mozambique may have been added accidentally during mounting of the specimen, and its locality is thus cited here with reservation]

16. **Euphorbia prostrata** Aiton, Hort. Kew. **2**: 139 (1789). —Boissier in de Candolle, Prodr. **15**, 2: 47 (1862). —Hiern, Cat. Afr. Pl. Welw. **1**: 942 (1900). —N.E. Brown in F.T.A. **6**, 1: 510 (1911); in F.C. **5**, 2: 245 (1915). —Robyns, Fl. Sperm. Parc Nat. Alb. **1**: 477 (1948). — Suessenguth & Merxmüller, [Contrib. Fl. Marandellas Distr.] Proc. & Trans. Rhod. Sci. Ass. **43**: 84 (1951). —Keay in F.W.T.A., ed. 2, **1**: 421 (1958). —Hadidi in Bull. Jard. Bot. Belg. **43**: 98 (1973). —S. Carter in F.T.E.A., Euphorbiaceae, part 2: 421, fig. 78/2 (1988). Type from West Indies.

Prostrate much-branched annual herb, the whole plant often tinged purplish; branches to 20 cm long, glabrous on the underside, pilose above with short curled hairs. Leaves with a petiole to 1 mm long; stipules free on the upper surface of the branch, 0.5 mm long, triangular, pilose, joined on the lower surface to 1 mm long, forming a broad triangle with 2 unequal teeth; lamina to 8 × 5 mm, ovate, rounded at the apex, obliquely rounded at the base, obscurely toothed at the margin, upper surface glabrous, lower surface sparsely pilose towards the apex. Cyathia terminal and pseudoaxillary on short leafy shoots, solitary on peduncles to 1.25 mm long, 1 × 0.6 mm, with barrel-shaped involucres; glands 4, minute, red, with minute white or pink appendages; lobes minute, triangular, pilose. Male flowers few: bracteoles hair-like; stamens 1 mm long. Female flower: ovary pedicellate; styles 0.2 mm long, spreading, bifid to the base. Capsule exserted on a reflexed pilose pedicel 1.5 mm long, 1.25 × 1.25 mm, acutely 3-lobed with a truncate base; the base and purple-tinged sutures beset with long spreading hairs. Seeds 1 × 0.5 mm, oblong-conical, acutely 4-angled, with numerous distinct transverse ridges and grooves, greyish-brown.

Botswana. N: Maun, fl. & fr. ii.1967, *Lambrecht* 44 (K; SRGH). SE: Mochudi, Phutodikobo Hill, fl. & fr. 10.iii.1967, *Mitchison* 24 (K). **Zambia**. B: Mongu, fl. & fr. 24.i.1960, *Gilges* 922 (NDO). N: Mbala Distr., Lunzua Power Station, fl. & fr. 22.v.1962, *Richards* 16488 (K; SRGH). C: Luangwa Valley Game Reserve, Mfuwe Camp, fl. & fr. 12.xii.1968, *Astle* 5391 (K; NDO; SRGH). E: Chipata, fl. & fr. 12.x.1967, *Mutimushi* 2318 (K; NDO; SRGH). S: Kalomo, fl. & fr. v.1909, *Rogers* 8224 (K; SRGH). **Zimbabwe**. N: Gokwe, fl. & fr. 10.xi.1963, *Bingham* 889 (SRGH). W: Hwange Distr., Gwampa Forest Reserve, fl. & fr. ii.1955, *Goldsmith* 120/55 (K; SRGH). C: Makoni Distr., near Maidstone, fl. & fr. 23.ii.1931, *Norlindh & Weimarck* 5142 (COI;

K; SRGH). E: Mutare, fl. & fr. 22.vi.1946, *Chase* 218 (BM; K; SRGH). S: Chiredzi, fl. & fr. 1.ii.1971, *P.E. Taylor* in *GHS* 217359 (K; SRGH). **Malawi**. N: Mzimba Distr., 5 km west of Mzuzu at Katoto, fl. & fr. 14.iv.1974, *Pawek* 8323 (K; MAL; MO; SRGH). C: Dedza Distr., Chongoni Forestry School, fl. & fr. 13.iii.1967, *Salubeni* 585 (K; MAL; SRGH). S: Blantyre Distr., Maone Estate, 2 km north of Limbe, fl. & fr. 17.iv.1970, *Brummitt* 9932 (K; MAL). **Mozambique**. N: Namapa Distr., Namapa, fl. & fr. 30.iii.1961, *Balsinhas & Marrime* 327 (BM; COI; K; SRGH). Z: Lugela Distr., Namagoa, fl. & fr. 6.v.1948, *Faulkner* 259 (K). MS: Gorongosa Distr., Parque Nacional da Gorongosa (Gorongosa Nat. Park), Urema Plains, fl. & fr. iv.1972, *Tinley* 2536 (SRGH). GI: Massinga Distr., 5 km south of Massinga, fl. & fr. ii.1938, *Gomes e Sousa* 2082 (COI; K). M: Maputo, fl. & fr. 22.ii.1920, *Borle* 342 (K).

A common weed of the tropics and subtropics originally introduced from the West Indies. Disturbed ground in gardens, on cultivated land and by roadsides, especially in dry sandy soils; 0–1650 m.

17. **Euphorbia mossambicensis** (Klotzsch & Garcke) Boiss. in de Candolle, Prodr. **15**, 2: 36 (1862), as "*mozambicensis*". —N.E. Brown in F.T.A. **6**, 1: 509 (1911), as "*mozambicensis*". — Eyles in Trans. Roy. Soc. South Africa **5**: 399 (1916), as "*mozambicensis*". —S. Carter in Kew Bull. **39**: 644 (1984); in F.T.E.A., Euphorbiaceae, part 2: 423, fig. 78/3 (1988). Type: Mozambique, Sena, *Peters* 33 (B†, holotype; K, fragment of holotype).

Anisophyllum mossambicense Klotzsch & Garcke in Klotzsch, Nat. Pflanzenk. Tric.: 30 (1860).

Euphorbia mozambicensis var. *nyasica* N.E. Br. in F.T.A. **6**, 1: 510 (1911). Syntypes: Malawi, Nyika Plateau, Nyamkowa (Nymkowa), fl. & fr. Feb. & Mar.1903, *McClounie* 169 (K); Mt. Malosa, fl. & fr. Nov. & Dec. 1896, *Whyte* s.n. (K).

A much-branched prostrate annual herb; branches up to 35 cm long with upper surface covered in short appressed hairs, the lower surface glabrous. Leaves with a petiole to 0.5 mm long; stipules to 0.4 mm long, linear, sometimes 2–3-toothed at the broader base; lamina to 14 × 8 mm, rounded and entire or obscurely toothed at the apex, very obliquely rounded at the base, glabrous or rarely with a few scattered hairs around the margin. Cyathia terminal and pseudoaxillary on short leafy shoots, solitary on peduncles 0.5–1 mm long, c. 1.25 × 1.25 mm, with cup-shaped involucres; glands 4, to almost 1 mm wide but usually much less, reddish with obvious white or pinkish lobed appendages to 0.6 × 1.5 mm; involucral lobes minute, triangular, with ciliate margins. Male flowers: bracteoles laciniate; stamens 1.25 mm long. Female flower: ovary pedicellate; styles very short (0.3 mm long), erect, bifid to halfway. Capsule exserted on a reflexed pedicel up to 3 mm long, 1.6 × 1.6 mm, acutely 3-lobed with truncate base, usually glabrous, occasionally with a few long spreading hairs. Seeds 0.9 × 0.5 mm, ovate-conical, 4-angled with a few very obscure ridges and grooves, pinkish-brown.

Botswana. N: Okavango Swamp, Xhamoga Lediba, fl. 26.i.1974, *P.A. Smith* 798 (SRGH). **Zambia**. C: 12 km east of Lusaka, fl. & fr. 5.i.1972, *Kornaś* 793 (K). S: Livingstone, fl. & fr. 4.xii.1962, *Bainbridge* 623 (NDO; SRGH). **Zimbabwe**. N: Hurungwe Distr., 13 km ESE of Chirundu Bridge, fl. & fr. 3.ii.1958, *Drummond* 5451 (BM; K; SRGH). W: Hwange National Park, 22.5 km WSW of Main Camp along Guvalala (Guvalalla) road, fl. & fr. 18.ii.1969, *Rushworth* 1547 (K; SRGH). S: Gwanda Distr., 38.5 km from Koodoovale Motel on road to Thuli (Tuli) Breeding Station by tributary of Mtshibizini R., fl. & fr. 19.iii.1959, *Drummond* 6042 (K; SRGH). **Malawi**. N: 40 km north of Chilumba, 32 km south of Karonga, fl. & fr. 14.iv.1976, *Pawek* 11027 (K; MAL; MO; SRGH; UC). C: Salima, road near Lake Nyasa Hotel, fl. & fr. 14.ii.1959, *Robson* 1598 (BM; K; MAL; SRGH). S: Zomba Distr., Mt. Malosa, fl. & fr. xi. & xii.1896, *Whyte* s.n. (K). **Mozambique**. Z: Macuri, fl. & fr. 1884–85, *Carvalho* s.n. (COI). T: Tete, fl. & fr. ii.1859, *Kirk* s.n. (K). MS: Caia Distr., Sena, fl. & fr. no date, *Peters* 33 (K, fragment of holotype).

Also in Tanzania. On sandy soils in grassland and open woodland; 60–1200 m.

The variation in pubescence, notably on the capsule, seems to be influenced only by environmental conditions. However, one specimen, *Leach* 13650 from Zimbabwe, 27 km north of Beitbridge, has a sparse pubescence of exceptionally long hairs around the stems and on the undersurface of the leaves. As the petaloid appendages on the involucral glands are also rather small, the specimen may eventually prove to represent a distinct taxon.

18. **Euphorbia schlechteri** Pax in Bot. Jahrb. Syst. **28**: 26 (1899). —N.E. Brown in F.C. **5**, 2: 247 (1915). Type: Mozambique, Ressano Garcia, *Schlechter* 11915 (COI; K; PRE).

Perennial herb with woody branches to 20 cm long, densely pilose with long soft hairs. Leaves with a petiole 1 mm long; stipules c. 0.5 mm long, linear from a broad base; lamina to 12 × 6 mm, ovate, rounded at the apex, obliquely subcordate at the base, dentate at the margin, somewhat leathery, upper surface glabrous, lower surface sparsely pilose. Cyathia terminal and pseudoaxillary, solitary on peduncles to 1 mm long, c. 1 × 1 mm with cup-shaped involucres, glabrous; glands 4, c. 0.5 mm wide, transversely elliptic, with white narrow minutely crenulate appendages; lobes acutely triangular, ciliate. Male flowers: bracteoles few, filamentous; stamens 1 mm long. Female flower: ovary pedicellate, densely pilose; styles 0.5 mm long, spreading, deeply bifid. Capsule exserted on a reflexed pedicel to 3.5 mm long, 2.5 × 2.5 mm, deeply 3-lobed with truncate base, pilose. Seeds 1.5 × 1.2 mm, ovoid, obtusely 4-angled, shallowly pitted, reddish-brown.

Mozambique. M: Ressano Garcia, fl. & fr. 24.xii.1897, *Schlechter* 11915 (COI; K).
Known only from the type collection; c. 300 m altitude, with no other data.
As no other material of this distinctive species has apparently ever been collected, it seems possible that Schlechter's plant may have been introduced. However, I have been unable to identify it with any species from elsewhere.

19. **Euphorbia zambesiana** Benth. in Hooker's Icon. Pl. **14**: t. 1305 (1880). —N.E. Brown in F.T.A. **6**, 1: 500 (1911). —Brenan in Mem. New York Bot. Gard. **9**: 66 (1954). —S. Carter in F.T.E.A., Euphorbiaceae, part 2: 428, fig. 78/5 (1988). Syntypes: Malawi, by Zomba Mt., fl. & fr. x.1861, *Meller* s.n. (K); Shire Highlands, Blantyre, fl. x.1878, *Buchanan* 10 (K).

Perennial herb with a woody twisted root c. 1 cm thick and up to 25 cm long or more, producing several woody underground stems to 3 cm long branching profusely at ground level, whole plant glabrous or densely pilose; branches densely rebranching, leafy and prostrate up to 15(35) cm long or more, floriferous and erect to c. 5 cm high, often tinged red. Leaves with a petiole to 2 mm long; stipules c. 1 mm long, linear or deeply divided into 2–4 linear teeth; lamina to 35 × 9 mm, lanceolate to ovate, minutely apiculate at the apex, obliquely subcordate at the base, entire or occasionally minutely toothed at the very narrow cartilaginous margin, lower surface often tinged red. Cyathia terminal and pseudoaxillary, solitary on peduncles 1–25 mm long, 2 × 2.5 mm with broadly cup-shaped involucres; glands 4 in bisexual cyathia, or sometimes 5 in cyathia which develop only male flowers, c. 1 mm wide, transversely elliptic, red with conspicuous white or pink appendages to 1.5 × 2 mm, entire or shallowly 2–3-lobed; involucral lobes 0.8 mm long, acutely triangular, margin ciliate. Male flowers many, especially in unisexual cyathia: bracteoles deeply laciniate, apices ciliate; stamens 2 mm long. Female flower: ovary pedicellate; styles 1 mm long, erect, bifid to halfway. Capsule exserted on a reflexed pedicel 5.5 mm long, 3 × 3 mm, deeply 3-lobed. Seeds 1.75 × 1.25 mm, ovate-conical, deeply pitted, reddish-buff.

Var **zambesiana** TAB. **73**, fig. C.
Euphorbia puggei Pax in Bot. Jahrb. Syst. **19**: 118 (1894). Type from Dem. Rep. Congo.

Plant entirely glabrous; leaves to 9 × 35 mm, lanceolate to ovate.

Zambia. N: Samfya, fl. & fr. 30.ix.1953, *Fanshawe* 320 (K; NDO). W: Kitwe Distr., near Nkana above the Buchi (Uchi) Dambo, fl. & fr. 25.ix.1947, *J.P.M. Brenan & R.A.F. Brenan* 7957 (EA; K; NDO). C: Mkushi Distr., c. 135 km south of Kanona on Kapiri Mposhi road, fl. & fr. 5.vii.1960, *Richards* 12829 (K; SRGH). **Malawi**. N: Chitipa Distr., Chisenga, foot of Mafinga Mts., fl. & fr. 8.xi.1958, *Robson & Fanshawe* 514 (BM; K; MAL). C: Dedza Distr., Chongoni Forest Reserve, Ciwawo (Chiwao) Hill, fl. & fr. 18.x.1967, *Salubeni* 853 (K; MAL; SRGH). S: Mulanje Distr., Phalombe Plain, fl. & fr. 14.viii.1960, *Leach* 10451 (K; SRGH). **Mozambique**. M: Namaacha Distr., Goba Fronteira, fl. 28.vi.1961, *Balsinhas* 493 (K).
Also in Tanzania, Rwanda and Burundi, southern Dem. Rep. Congo and Angola. Pyrophyte of open miombo, wooded grassland and montane grassland, often on laterite; 400–2150 m.
This is one of the first species to appear after burning, producing short, erect, very floriferous shoots with a few scale-like leaves at the underground base, later giving way to longer, more leafy, prostrate branches. Amongst the first shoots are solitary peduncles often bearing unisexual (male), rarely bisexual, cyathia with 5 glands.

Var. **villosula** (Pax) N.E. Br. in F.T.A. **6**, 1: 501 (1911). —S. Carter in Kew Bull. **39**: 647 (1984); in F.T.E.A., Euphorbiaceae, part 2: 429 (1988). Syntypes from Tanzania.
Euphorbia villosula Pax in Bot. Jahrb. Syst. **19**: 118 (1894).

Euphorbia angolensis Pax in Bot. Jahrb. Syst. **19**: 117 (1894). Type from Angola.
Euphorbia poggei var. *benguelensis* Pax in Bot. Jahrb. Syst. **23**: 532 (1897). Type from Angola.
Euphorbia poggei var. *villosa* Pax in Bull. Herb. Boissier **6**: 737 (1898). Type from Angola.
Euphorbia serpicula Hiern, Cat. Afr. Pl. Welw. **1**: 941 (1900). Type from Angola.
Euphorbia andongensis Hiern, Cat. Afr. Pl. Welw. **1**: 943 (1900). Type from Angola.
Euphorbia zambesiana var. *benguelensis* (Pax) N.E. Br. in F.T.A. **6**, 1: 501 (1911).

As for var. *zambesiana*, but the whole plant, including the capsule, densely pilose with long spreading hairs, or at least some hairs present on stems, leaves or capsules. Leaves to 12 × 7 mm, ovate to ovate-lanceolate.

Zambia. N: Mbala Distr., far side of Nkali Dambo, fl. 21.viii.1956, *Richards* 5896 (K; SRGH).
Also in Uganda, west and south Tanzania, Rwanda and Burundi, southern Dem. Rep. Congo and Angola. In open woodland; 900–1650 m.
Distribution of this pilose form overlaps that of the typical variety, but so far has not been collected east of northern Zambia. The degree of hairiness is extremely variable, from all parts of the plant being densely hairy to the stems only or sometimes also the leaves and/or the capsules with a few sparse hairs. All these forms occur with each other and with the glabrous form, sometimes within the same gathering, so no justification can be found for upholding other species or varieties. A particularly hairy form is common in the Mbala District of Zambia, extending into southwest Tanzania.

20. **Euphorbia rubriflora** N.E. Br. in F.T.A. **6**, 1: 509 (1911). —Eyles in Trans. Roy. Soc. South Africa **5**: 400 (1916). —Wild in Clark, Victoria Falls Handb.: 148 (1952). Syntypes: Zambia, Livingstone, fl. & fr. v.1909, *Rogers* 7132 (K); Zimbabwe, Victoria Falls, fl. & fr. i.1906, *Allen* 264 (K; SRGH, isosyntype).
    Euphorbia sp. of Eyles in Trans. Roy. Soc. South Africa **5**: 400 (1916) pro parte as to *Allen* 264.

Annual or short-lived perennial herb with a woody root up to 5 mm in diameter; branches prostrate, to 25 cm long, pubescent on the upper surface with short crisped hairs. Leaves with a petiole to 1 mm long; stipules linear, to 1.5 mm long; lamina to 13 × 8 mm, ovate, rounded apex, obliquely subcordate at the base, entire at the margin, glabrous. Cymes terminal and pseudoaxillary, clustered into short leafy shoots. Cyathia solitary on peduncles 0.5 mm long, c. 1.8 × 1.8 mm, with sparsely pilose cup-shaped involucres; glands 4, c. 0.7 mm wide, transversely elliptic, with conspicuous white, pink or red appendages, 2 of these c. 1 × 1.5 mm, the other 2, on either side of the gap through which the capsule is exserted, extended obliquely to c. 3 × 1.5 mm; lobes 0.5 mm long, acutely triangular, ciliate. Male flowers: bracteoles few, filiform; stamens many, 2.5 mm long. Female flower: ovary pedicellate; styles 2 mm long, erect, free to the base. Capsule just exserted on a reflexed pedicel to 2 mm long, 1.8 × 1.8 mm, acutely 3-lobed with truncate base, pilose on the angles. Seeds 1.3 × 0.7 mm, oblong-conical, sharply 4-angled, minutely pitted and wrinkled, pinkish-grey.

Botswana. N: Chobe Distr., Kasane, fl. & fr. 20.ii.1968, *Mutakela* 178 (SRGH). **Zambia**. S: Livingstone, fl. & fr. iv.1909, *Rogers* 7098 (SRGH). **Zimbabwe**. W: Hwange Distr., above Deka, fl. & fr. iii.1918, *Eyles* 1290 (BM; SRGH).
A species of limited distribution, concentrated in the Hwange District area. Dry exposed often stony soils, in sparse grassland; 700–1000 m.

Sect. 3. ARTHROTHAMNUS (Klotzsch & Garcke) Boiss. in de Candolle, Prodr. **15**, 2: 74 (1862).
Arthrothamnus Klotzsch & Garcke in Klotzsch, Nat. Pflanzenk. Tric.: 62 (1860).

Monoecious or often dioecious woody-based shrubs, with a reduced main stem; branches succulent to semi-woody, strictly cylindrical. Leaves (bracts) opposite, small and often scale-like, quickly deciduous leaving a calloused scar; stipules apparently absent or minute and glandular. Cymes branching dichotomously many times, with rays progressively shorter above; bracts similar to the leaves, deciduous. Cyathia very small; involucres bisexual, glands 5, entire, without appendages. Stamens just exserted, with subsessile anthers. Perianth often obvious below the ovary, 3-lobed. Capsule subsessile. Seeds conical, obtusely 4-angled, ecarunculate.

21. **Euphorbia rectirama** N.E. Br. in F.C. **5**, 2: 283 (1915). —White, Dyer & Sloane, Succ. Euphorb. **1**: 179 (1941). Types from South Africa.

Dioecious shrubs to 1 m high, densely branched from the base; branches opposite, succulent, terete, c. 3 mm thick. Leaves with a broad petiole c. 0.5 mm long, quickly deciduous, leaving a prominent calloused brown scar; stipules absent; lamina fleshy, to 3 × 1.5 mm, ovate. Cymes axillary and terminal, each forking several times with primary rays 1–2.5 cm long. Bracts c. 2 × 1.2 mm, spathulate, minutely ciliolate on the margins at the base. Cyathia c. 2 × 2.5 mm, with cup-shaped involucres, subsessile with a minutely ciliolate pedicel; glands 5, c. 0.5 × 1 mm, transversely elliptic, spreading; lobes 0.5 mm long, rounded, margin toothed, ciliolate. Male involucres: bracteoles linear, ciliate; stamens 1.8 mm long. Female involucres: a few bracteoles present; perianth obvious, 3-lobed; ovary glabrous; styles 1 mm long, joined at the base, with spreading bifid apices. Capsule just exserted on an erect pedicel 1.5–2 mm long, 3.2 × 3.5 mm, deeply 3-lobed. Seeds 2.5 × 1.5 mm, conical, obtusely 4-angled, minutely verrucose, grey.

Botswana. SW: Kgalagadi Distr., Tshabong, fl. (male & female), 25.ii.1960, *Yalala* 88 (K; SRGH). SE: Kweneng Distr., Gabane Hills, fl. (male), 2.iv.1977, *O.J. Hansen* 3116 (GAB; K; PRE; SRGH).
Occurs commonly in South Africa (Cape Province and western Transvaal). In sandy soils amongst rocks, often on kopjes; c. 800 m.

Sect. 4. POINSETTIA (Graham) Baill., Étude Gén. Euph.: 284 (1858).
Poinsettia Graham in Edinburgh New Philos. J. **20**: 412 (1836).

Erect annual herbs, or large shrubs with semi-succulent branches. Cyathia in densely branching cymes, with basal bracts large and leaf-like, sometimes brightly coloured. Involucres bisexual, with 1 funnel-shaped gland and 5 lobes. Stamens just exserted, with subsessile anthers, and bracteoles included in the involucre. Perianth-rim of the female flower sometimes prominent; styles joined at the base, with bifid apices. Capsule exserted on a reflexed pedicel. Seeds conical, verrucose (with warts), without a caruncle.

1. Robust shrub to 4 m high, with stout semi-succulent branches; floral bracts brilliant red
   · 24. *pulcherrima*
 – Herbs to 1 m high; floral bracts uniformly green or with a basal red blotch · · · · · · · · · 2
2. Floral bracts ovate (in tropical Africa), uniformly green; opening of involucral gland
   circular; warts on seed surface rounded · 22. *heterophylla*
 – Floral bracts panduriform, with a red blotch at the base; opening of involucral gland
   elliptical; warts on seed surface pointed · 23. *cyathophora*

22. **Euphorbia heterophylla** L., Sp. Pl.: 453 (1753). —Keay in F.W.T.A., ed. 2, **1**: 421 (1958). — S. Carter in F.T.E.A., Euphorbiaceae, part 2: 431 (1988). TAB. **74**, fig. A. Type from tropical America.
   Euphorbia geniculata Ortega, Hort. Matr. Dec.: 18 (1797). —Agnew, Upland Kenya Wild Fl.: 222 (1974). Type from Cuba.

Annual herb erect to 1 m high, often tinged red; stem hollow; branches sparse, glabrous to sparsely pilose towards the apices. Leaves with a petiole to 2(4) cm long; stipules modified as fairly conspicuous purplish glands; lamina to 12 × 6 cm, ovate, obtuse at the apex, cuneate at the base, with minute distant gland-tipped teeth at the margin, occasionally more coarsely toothed, glabrous to sparsely pilose around the edges on the upper surface, pilose with septate hairs especially on the midrib and nerves on the lower surface, glabrescent. Cymes terminal and axillary, each forking c. 5 times, with rays progressively shorter from c. 15 cm to c. 2 mm long and cyathia densely clustered. Basal bracts similar to the leaves but paler green, progressively smaller, more lanceolate and sessile above. Cyathia c. 3.5 × 2.5 mm with barrel-shaped involucres, glabrous; gland peltate, 1 mm long, funnel-shaped, the opening circular, 1.2 mm across, often red-rimmed; lobes c. 1.3 mm long, subcircular, deeply and sharply toothed with margins minutely ciliate. Male flowers: bracteoles few, ligulate, feathery; stamens 4 mm long. Female flower: ovary pedicellate, glabrous or

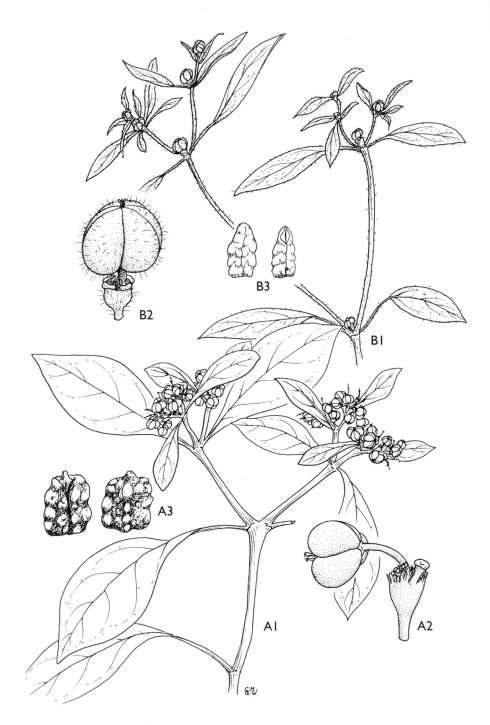

Tab. 74. A. —EUPHORBIA HETEROPHYLLA. A1, fruiting branch (× ²/₃); A2, cyathium with capsule (× 4); A3, seeds (× 6), A1–A3 from *Norrgrann* 327. B. —EUPHORBIA BENTHAMII. B1, fertile branch (× ²/₃), from *Richards* 1500; B2, cyathium with capsule (× 4); B3, seeds (× 4), B2 & B3 from *Fanshawe* 6311. Drawn by Christine Grey-Wilson.

occasionally with minute scattered hairs, the perianth forming an obvious rim; styles c. 1 mm long, occasionally minutely puberulous, bifid to halfway. Capsule exserted on a reflexed pedicel to 6 mm long, c. 4.5 × 5.5 mm, deeply 3-lobed. Seeds 2.6 × 2.4 mm, conical with acute apex, bluntly verrucose, blackish-brown.

Botswana. N: Kwando, Hunters Camp, fl. 26.i.1978, *P.A. Smith* 2251 (K; PRE; SRGH). **Zambia**. W: Kitwe, fl. & fr. 18.i.1960, *Fanshawe* 5356 (K). C: Mt. Makulu Agricultural Research Station, 19 km south of Lusaka, fl. & fr. 7.vi.1956, *Angus* 1325 (K; SRGH). S: Gwembe Distr., Siavonga, fl. & fr. 15.x.1972, *Kornas* 2384 (K). **Zimbabwe**. N: Makuti, Tsetse Control Centre, fl. & fr. 17.ii.1981, *Philcox & Leppard* 8686 (K; SRGH). W: Matobo Distr., Hope Fountain Mission, fl. & fr. 14.iv.1973, *Norrgrann* 327 (K; PRE; SRGH). C: Chegutu (Hartley), fl. & fr. 21.ix.1968, *R.M. Hornby* 3461 (K; SRGH). E: Mutare (Umtali), fl. & fr. 7.i.1956, *Chase* 5946 (K; SRGH). S: Masvingo (Fort Victoria), fl. & fr. 15.iii.1967, *Rushworth* 324 (SRGH). **Malawi**. N: Nkhata Bay, shore of Lake Malawi (Nyassa) fl. & fr. 20.ii.1961, *Richards* 14419 (K). C: Salima Distr., Chipoka, fl. & fr. 13.iii.1985, *Salubeni & Kaunda* 4108 (MAL; SRGH). S: Zomba, fl. & fr. 15.xii.1979, *Salubeni & Tawakali* 2682 (MAL; SRGH). **Mozambique**. Z: Lugela Distr., Missão Munguluni (M'guluni Mission), fl. & fr. i.1947, *Faulkner* Kew 132 (K). MS: Gondola, fl. & fr. ii.1922, *Honey* 701 (K; PRE). M: Namaacha Distr., Goba Fronteira, fl. & fr. 11.i.1980, *de Koning* 7991 (K).

A pantropical weed originating from Central America and occurring throughout tropical Africa. A weed of cultivation; 400–1400 m.

Plants occurring in tropical Africa show little of the variation which characterises this very widespread species, with scarcely any indication of the irregularity in leaf shape giving the species its name. The leaves, especially the upper ones, are occasionally more lanceolate or slightly panduriform in shape, and sometimes a pale cream or purplish blotch occurs at the base of the bracts.

23. **Euphorbia cyathophora** Murray in Commentat. Soc. Regiae Sci. Gott. **7**: 81, t. 1 (1786). — S. Carter in F.T.E.A., Euphorbiaceae, part 2: 431 (1988). Type from tropical America.

Shrubby annual herb, erect to 1 m high, with hollow glabrous stems and branches. Leaves with a petiole to 1.5 cm long, pilose; stipules modified as brownish glands; lamina to 10 × 5 cm, markedly panduriform, obtuse at the apex, cuneate at the base, shallowly and irregularly toothed at the margin, glabrous on upper surface, sparsely pilose with septate hairs on lower surface, glabrescent. Cymes terminal, each forking c. 4 times, with rays progressively shorter from 5 cm to 2 mm long and cyathia densely clustered. Basal bracts similar to the leaves but with a bright orange-red blotch at the base, the upper bracts progressively smaller, more lanceolate, subsessile and entirely red. Cyathia c. 3.5 × 3 mm with barrel-shaped involucres, glabrous; gland peltate, 1.5 mm long, funnel-shaped, the opening transversely elliptic, 2 mm wide; lobes 1.5 × 1.5 mm, rounded, margin deeply and bluntly toothed. Male flowers: bracteoles ligulate, feathery; stamens 4 mm long. Female flower: ovary pedicellate, glabrous; styles 2 mm long, bifid almost to the base. Capsule exserted on a reflexed pedicel to 5.5 mm long, 4 × 5 mm, deeply 3-lobed. Seeds 2.8 × 2.2 mm, ovoid-conical with acute apex and warts sharply pointed, blackish-brown.

Botswana. N: Chobe Distr., Kasane, fl. & fr. iv.1966, *Mutakela* 4/66/10 (SRGH). **Zambia**. W: Kitwe, fl. & fr. 4.ii.1964, *Mutimushi* 557 (K; SRGH). C: Lusaka, Handsworth Park, fl. & fr. 21.i.1962, *Best* 312 (SRGH). **Zimbabwe**. W: Victoria Falls Village, fl. & fr. 6.iii.1974, *Gonde* 47/74 (K; PRE; SRGH). C: Harare, public park, fl. & fr. 6.xi.1972, *Biegel* 4108 (K; PRE; SRGH). E: Mutare (Umtali), fl. & fr. 7.i.1956, *Chase* 5948 (K; PRE; SRGH). S: Bikita Distr., Umkondo Mine, fl. & fr. 21.v.1972, *Leach & Cannell* 14892 (K; SRGH). **Mozambique**. GI: Bilene Distr., 1 km from Chissano on road to Maguenza, fl. & fr. 13.ii.1959, *Barbosa & Lemos* 8402 (COI; K; PRE). M: Maputo, Costa do Sol, fl. & fr. ix.1966, *Charneca* 129 (COI).

Pantropical, introduced into a number of places throughout tropical Africa and becoming naturalised. A garden escape around habitation; 0–1500 m.

24. **Euphorbia pulcherrima** Willd. ex Klotzsch in Allg. Gartenzeitung **2**: 27 (1834). —F. White, F.F.N.R.: 199 (1962). Type from Central America.
    Poinsettia pulcherrima (Klotzsch) Graham in Edinburgh New Philos. J. **20**: 412 (1836).

Robust shrub to 4 m high, branching from the base; branches stout, softly woody, with pronounced leaf scars. Leaves with a petiole to 5 cm long; stipules modified as conspicuous purplish-brown glands; lamina to c. 20 × 12 cm, ovate or often panduriform, attenuate with rounded tip at the apex, cuneate at the base, pubescent with short crisped hairs especially on the lower surface, dark green above, paler beneath. Cymes terminal, pseudumbellate with 3(4) rays 1–3 cm long, each 3–6-forked. Bracts leaf-like, with a petiole 1–2 cm long, lamina to c. 15 × 4 cm, progressively smaller above, lanceolate, glabrescent, brilliant red. Cyathia 8 × 7 mm, with deeply cup-shaped involucres, glabrous; gland peltate, 3 mm long, funnel-shaped with the opening transversely elliptic 5 mm wide, yellow-rimmed; lobes 5, 4 mm wide, with large irregular rounded marginal teeth. Male flowers: bracteoles 6 mm long, laciniate, pubescent; stamens very many, 8 mm long. Female flower: ovary glabrous, exserted on a reflexed pedicel 1.5 cm or more long; perianth obvious as a thickened rim below the ovary; styles 3.5 mm long, joined for 1.5 mm with spreading deeply bifid apices. Capsule and seeds not seen.

Malawi. S: Zomba Distr., lower slopes of Zomba Plateau by Mponda Stream, fl. 26.iv.1980, *Brummitt, Masiye & Tawakali* 15563 (K; MAL).
   Recorded as an established garden escape apparently in this one instance only; 1140 m.
   An ornamental shrub often cultivated in gardens in the tropics or as a pot-plant elsewhere. Cultivars have been selected, with bracts in varying shades of red, pink and yellow.

Sect. 5. PSEUDACALYPHA Boiss. in de Candolle, Prodr. **15**, 2: 98 (1862).
—S. Carter in F.T.E.A., Euphorbiaceae, part 2: 442 (1988).

Erect, branching annual herbs (no perennials in the Flora Zambesiaca area), with stems sometimes woody or ± fleshy. Leaves petiolate; stipules modified as glands. Cymes axillary and in 3-branched terminal umbels, forking dichotomously many times but often with only one ray developing at each fork. Bracts similar to the leaves. Cyathia with bisexual involucres and 4 entire, spreading involucral glands. Stamens just exserted from the involucre. Capsule subglobose with a truncate base, scarcely exserted from the involucre. Seeds conical ornamented with pointed apex, 4-angled, ornamented with minute warts, without a caruncle.

1. Whole plant pilose with long spreading hairs; styles erect in a distinct column, to 2.5 mm long · 27. *crotonoides*
–   Plant sparsely pilose; styles spreading, to 1.5 mm long · 2
2. Capsule with short appressed hairs, and longitudinal ridges each side of the sutures · 25. *systyloides* subsp. *porcaticapsa*
–   Capsule with long spreading hairs, longitudinal ridges absent · · · · · · · · · · · 26. *benthamii*

25. **Euphorbia systyloides** Pax in Bot. Jahrb. Syst. **19**: 121 (1894). —N.E. Brown in F.T.A. **6**, 1: 520 (1911). —S. Carter in F.T.E.A., Euphorbiaceae, part 2: 443, fig. 82/4–6 (1988). Type from Zanzibar Island.
   Euphorbia holstii Pax in Bot. Jahrb. Syst. **19**: 121 (1894). Type from Tanzania.
   Euphorbia volkensii Pax in Engler, Pflanzenw. Ost-Afrikas **C**: 242 (1895). Type from Tanzania.

Subsp. **porcaticapsa** S. Carter in Kew Bull. **45**: 336 (1990). Type: Zimbabwe, Hurungwe Distr., Zambezi Valley, Rifa R., 520 m, fl. & fr. 24.ii.1953, *Wild* 4085 (K, holotype; EA; SRGH).

Annual herb to 1 m high, usually less, stem often woody at the base, the whole plant sparsely covered with long spreading crisped hairs. Leaves with a petiole to 4.5 cm long; glandular stipules purplish; lamina to 13 × 3.5 cm, lanceolate to ovate-lanceolate, acute at the apex, cuneate at the base, finely serrate with small gland-tipped teeth at the margin, with midrib prominent on the lower surface. Cymes axillary and in 3-branched umbels, with primary rays to 15 cm long, each forking up to 10 times. Bracts leaf-like, the upper ones subsessile. Cyathia sessile, 2.5 × 3.5 mm, with cup-shaped involucres, pubescent; glands 4, 1 × 1.5 mm, transversely elliptic, green; lobes 0.7 mm wide, rounded, margin finely ciliate. Male flowers: bracteoles fan-shaped, deeply laciniate, with feathery tips; stamens 2.5 mm long. Female flower: ovary densely hairy; styles c. 1 mm long, joined at the base, spreading, with somewhat flattened grooved apices. Capsule scarcely exserted, 5 × 6.5 mm, subglobose with truncate base and with a longitudinal ridge each side of the sutures (6 in all) which are fleshy on immature fruit, pubescent with short appressed hairs. Seeds 4 × 3 mm, conical, with acute apex, sharply 4-angled, with shallow warts in obscure horizontal lines, greyish-black.

Zambia. S: Gwembe Valley, fl. & fr. 10.vi.1963, *Lawton* 1097 (K). **Zimbabwe**. N: Mount Darwin Distr., Muzarabani C.L. (Mzarabani Tribal Trust Land), Masingwa (Musingwa) R., fl. & fr. 30.iv.1972, *Mavi* 1350 (K; SRGH). S: Bikita Distr., Moodie's Pass, fl. & fr. 24.v.1959, *Noel* 3887 (SRGH). **Malawi**. S: Mangochi Distr., Monkey Bay, fl. & fr. 17.iii.1985, *Salubeni & Kaunda* 4165 (MAL). **Mozambique**. Z: 50 km from Mocuba towards Namabida, fl. & fr. 2.vi.1949, *Barbosa & Carvalho* 2962 (K).

Not known elsewhere. In open woodland, often near streams; 400–1000 m.

Subsp. *systyloides* has a scattered distribution in East Africa with a concentration in northeast Tanzania. It is a taller plant, to 1.5 m high, with shorter, smaller leaves having shorter petioles, with smaller capsules without the fleshy longitudinal ridges typical of the southern subspecies, and seeds which are not sharply angled.

26. **Euphorbia benthamii** Hiern, Cat. Afr. Pl. Welw. **1**: 943 (1900). —N.E. Brown in F.T.A. **6**, 1: 518 (1911). —S. Carter in F.T.E.A., Euphorbiaceae, part 2: 445 (1988). TAB. **74**, fig. B. Type from Angola.

Annual herb to 1 m high, stem often woody at the base, the whole plant sparsely covered with long spreading white hairs. Leaves with a petiole to 2.5 cm long; glandular stipules purplish; lamina to 11.5 × 3 cm, acute at the apex, cuneate at the base, subentire to serrate with minute gland-tipped teeth at the margin, midrib prominent on the lower surface. Cymes axillary and in 3-branched umbels forking up to 15 times and spreading horizontally, with primary rays to 12 cm long. Bracts subsessile, leaf-like, usually reflexed. Cyathia sessile, 2.5 × 4 mm with cup-shaped involucres, pilose; glands 4, 1 × 1.5 mm, transversely elliptic, green becoming reddish-brown; lobes 0.7 mm in diameter, rounded, ciliate. Male flowers: bracteoles fan-shaped, feathery; stamens 2.5 mm long. Female flower: ovary densely hairy; styles 1 mm long, joined at the base, spreading, with thickened grooved apices. Capsule barely exserted on a pilose pedicel c. 2 mm long, 4 × 5 mm, subglobose with truncate base, pilose with long spreading hairs. Seeds 3 × 2 mm, conical with an acute apex, 4-angled, surface with irregular warts in 3–4 horizontal ridges, brownish-black.

Caprivi Strip. 32 km from Singalamwe on road to Katima Mulilo, fl. & fr. 3.i.1959, *Killick & Leistner* 3265 (K). **Zambia**. B: Mongu–Lealui, fl. & fr. 20.i.1966, *E.A. Robinson* 6802 (K; SRGH). S: Mazabuka Distr., Kafue Gorge, below Dam, fl. & fr. 30.xii.1971, *Kornaś* 738 (K). N: Mbala Distr., Itimbwe Gorge, fl. & fr. 6.iii.1960, *Richards* 12709 (EA; K; SRGH). C: Chisamba, fl. & fr. 28.ii.1966, *Fanshawe* 9595 (K; NDO). **Zimbabwe**. C: Chikomba Distr., between Chivhu (Enkeldoorn) and The Range, fl. & fr. 20.iv.1969, *Biegel* 2924 (K; SRGH). E: Mutare Distr., Laurance Ville, Vumba, fl. & fr. 3.iv.1955, *Chase* 5539 (COI; K; PRE; SRGH). S: Mberengwa Distr., Mnene Mission, 23.xi.1969, *Cannell* 87 (K; SRGH). **Malawi**. C: Ntchisi Forest Reserve, below Rest House, fl. & fr. 27.iii.1970, *Brummitt* 9451 (K; SRGH).

Also in southern Tanzania, Namibia and Angola. In mixed deciduous woodland, often near streams; 950–1500 m.

27. **Euphorbia crotonoides** Boiss. in de Candolle, Prodr. **15**, 2: 98 (1862). —N.E. Brown in F.T.A. **6**, 1: 518 (1911). —Eyles in Trans. Roy. Soc. South Africa **5**: 399 (1916). —S. Carter in F.T.E.A., Euphorbiaceae, part 2: 446, fig. 82/1–3 (1988). Type from Sudan.

Subsp. **crotonoides**
Euphorbia holstii var. *hebecarpa* Pax in Bot. Jahrb. Syst. **34**: 374 (1904). Syntypes from Kenya.
Euphorbia systyloides var. *hebecarpa* (Pax) N.E. Br. in F.T.A. **6**, 1: 521 (1911).

Annual herb to 1(1.5) m high, much-branched and somewhat fleshy, the whole plant covered in long spreading white hairs; stem often woody below; branches and upper part of the stem longitudinally ridged to distinctly winged. Leaves reflexed, with a 2-winged petiole to 2.5 cm long; glandular stipules dark red; lamina 3–9 × 0.5–5 cm, ovate to ovate-lanceolate, acute at the apex, base tapering into the petiole, serrate at the margin, often markedly so, with gland-tipped sometimes red-tinged teeth, midrib winged on the lower surface. Cymes axillary and in 3-branched umbels forking many times, with primary rays to 10 cm long. Bracts narrower than the leaves, with petioles to 5 mm long. Cyathia subsessile, 2.5 × 3.5 mm, with cup-shaped involucres; glands 4, 0.8 × 1.5 mm, transversely elliptic, yellow turning to red; lobes 1.75 mm wide, quadrangular, margin ciliate. Male flowers: bracteoles deeply laciniate, with apices ciliate; stamens 3 mm long. Female flower: ovary densely pilose; styles 2–2.5 mm long, puberulous, joined to almost halfway, erect to form a distinct column, with spreading thickened apices. Capsule exserted on a slightly curved pilose pedicel to 3 mm long, subglobose, 6.5–7 mm in diameter, with truncate base and shallow longitudinal grooves along the sutures, pilose with long spreading hairs. Seeds c. 4.5 × 3 mm, ovoid or conical, 4-angled, apex rounded or acute, surface obscurely to distinctly warted in irregular horizontal lines, brown to grey or reddish-black.

Botswana. N: 77 km north of Aha Mts., fl. & fr. 13.iii.1965, *Wild & Drummond* 6995 (K; PRE; SRGH). SW: western Kgalagadi, northern Etsha road, fl. & fr. 19.iii.1976, *Vahrmeijer Stee* 3147 (K; PRE). **Zambia**. B: Masese, fl. & fr. 12.iii.1960, *Fanshawe* 5470 (K; NDO). C: Lusaka, fl. & fr. 6.iii.1971, *Fanshawe* 11177 (NDO). S: Choma Distr., 6 km south of Mapanza, fl. & fr. 20.iii.1955, *E.A. Robinson* 1131 (K). **Zimbabwe**. N: east of Tashinga Camp, Matusadona National Park, fl. & fr. 28.iv.1976, *Mushori* 35 (SRGH). W: Hwange Distr., Dete (Dett), fl. & fr. 17.ii.1956, *Wild* 4773 (COI; K; LISC; SRGH). **Malawi**. N: Karonga Distr., Ngara, 41 km south of Karonga, fl. & fr. 25.iv.1975, *Pawek* 9553 (K; MAL; MO; PRE; SRGH).

Widespread from eastern Sudan, Ethiopia and East Africa southwards to South Africa (Transvaal) and Namibia. Usually on sandy soils in open woodland or on disturbed ground; 460–1250 m.

In the southern part of its distribution, including the Flora Zambesiaca area, growth of this widespread species is usually more lush, producing larger more ovate leaves, although those on the flowering branches (bracts) remain lanceolate. The size of the glandular teeth on the leaf-margins varies considerably, as does the density of the pubescence and colour of the seeds from brown to black.

Subsp. *narokensis* S. Carter, endemic in southwest Kenya, is a stunted herb with narrowly lanceolate leaves and obscurely angled, shallowly tuberculate seeds.

Sect. 6. EREMOPHYTON Boiss. in de Candolle, Prodr. **15**, 2: 70 (1862).
        —S. Carter in F.T.E.A., Euphorbiaceae, part 2: 448 (1988).

Erect, branching annual or perennial herbs, with stems often woody-based. Leaves petiolate; stipules minute, often deciduous. Cymes axillary and in terminal 3-branched umbels, forking dichotomously many times (in the Flora Zambesiaca area). Bracts leaf-like. Cyathia with bisexual involucres and 4 entire spreading involucral glands. Stamens just exserted from the involucre. Capsule oblong, well exserted from the involucre on a reflexed pedicel. Seeds compressed dorsiventrally, oblong and wrinkled in usually vertical lines, with a cap-like caruncle.

28. **Euphorbia pfeilii** Pax in Bot. Jahrb. Syst. **23**: 534 (1897). —N.E. Brown in F.C. **5**, 2: 250 (1915). Type from Namibia.

  Euphorbia glaucella Pax in Bull. Herb. Boissier **6**: 737 (1898). —N.E. Brown in F.T.A. **6**, 1: 514 (1911); in F.C. **5**, 2: 254 (1915). —Bremekamp & Obermeyer [Scientific Results of the Vernay-Lang Kalahari Expedition, March to September, 1930], in Ann. Transvaal Mus. **16**: 421 (1935). Syntypes from Namibia.

  Euphorbia anomala Pax in Bull. Herb. Boissier Ser. 2, **8**: 636 (1908) non Boissier (1862). Type from Namibia.

  Euphorbia kwebensis N.E. Br. in Bull. Misc. Inform., Kew **1909**: 137 (1909). Syntypes: Botswana, Kgwebe Hills, 7.i.1897, *Maj. Lugard* 143 (K); and 1.i.1898, *Mrs. Lugard* 81 (K).

Annual or short-lived perennial herbs to 60 cm high, glabrous, with a main stem to 15 cm high, woody-based; branches with prominent calloused scars from fallen leaves and cyathia. Leaves (mostly floral bracts) with a petiole to 8 mm long; stipules minute, subulate, soon deciduous and evident only on young growth; lamina to 5.5 × 1 cm, usually much less, linear to linear-lanceolate, rounded at the apex, cuneate at the base, usually with a few minute cartilaginous teeth at the margin especially towards the apex. Cymes axillary and in 3-branched umbels forking many times, with primary rays to 15 cm long. Bracts leaf-like. Cyathia subsessile, 1.7 × 2 mm, with cup-shaped involucres, glabrous or with a few very short appressed hairs around the top; glands 4, 0.8 × 0.5 mm, transversely elliptic with a thin paler rim on the outer margin; lobes 0.5 mm long, triangular, ciliate. Male flowers: bracteoles few, strap-shaped, ciliate; stamens 1.5 mm long. Female flower: ovary glabrous or pubescent; styles 1 mm long, joined at the base, with apices bifid to halfway. Capsule exserted on a curved pedicel to 2.5 mm long, 3.5 × 3 mm, oblong, obtusely 3-lobed, sparsely covered with short appressed hairs especially on the upper half, rarely entirely glabrous. Seeds 2.3 × 1.8 mm, oblong-rectangular, dorsally flattened, surface distinctly verrucose, grey; caruncle 1 mm across, pale brown.

Botswana. N: Kgwebe Hills, fl. & fr. i.1897, *Lugard* 143 (K). **Zimbabwe**. S: Beitbridge Distr., Mtetengwe R., fl. & fr. 26.ii.1967, *Leach* 13649 (K; SRGH).

Apparently limited in distribution in Zimbabwe to the Beitbridge District, while the syntypes of *E. kwebensis* from the Kgwebe Hills are the only specimens seen from Botswana. It also occurs in the Zoutpansberg mountains of South Africa, and extensively in Namibia. One specimen was seen from southern Angola. In hot dry areas in exposed gravelly sandy soils and rocky slopes; c. 500–1000 m.

With its short internodes, swollen nodes and denticulate ovate floral bracts, I cannot agree with Merxmüller (Prodr. Fl. SW. Afrika, fam. part 67: 25 (1967)) that *E. glanduligera* Pax from Namibia should be identified with this species. But *E. pfeilii* and *E. glaucella* are undoubtedly synonymous, their types and original descriptions representing leaf-forms from dry and damp situations respectively.

Sect. 7. ESULA Dumort., Fl. Belg.: 87 (1827).

Annual or perennial herbs, erect and sometimes shrubby. Leaves sessile or subsessile, usually entire; stipules absent. Cyathia in terminal and often axillary umbellate cymes; bracts deltoid, or leaf-like below the umbel. Involucres bisexual, or the primary central one of the umbel, if present, entirely male; glands 4, rarely 5–8 on involucres with all male flowers, entire or more often with 2 horns; lobes 5. Stamens with anthers clearly exserted from the involucre, but bracteoles included. Perianth of female flower reduced to a rim below the ovary; styles usually deeply bifid. Capsule exserted on a reflexed pedicel. Seeds ovoid, with a caruncle.

1. Capsules with fleshy ridges along the angles; seeds with rows of deep pits · · · · · 29. *peplus*
 - Capsules smooth or verrucose; seeds smooth · 2
2. Capsules smooth · 3
 - Capsules verrucose · 10
3. Leaves lanceolate, 2.5 × 1 cm to 15 × 2 cm · 4
 - Leaves linear to linear-lanceolate, 2 × 1 mm to 40 × 5 mm · · · · · · · · · · · · · · · · · · · 7
4. Herbs with simple or sparsely-branched stems to 30 cm high · · · · · · · · · · · · 36. *daviesii*
 - Branching shrubby herbs 0.5–2(4.5) m high · 5
5. Branches minutely pubescent; floral bracts yellow · · · · · · · · · · · · · · · · · · · 31. *citrina*
 - Branches glabrous or pubescent with long crisped hairs; floral bracts green · · · · · · · · ·6
6. Cymes forking many times; involucral glands horned · · · · · · · · · · · · · · 30. *schimperiana*
 - Cymes unbranched; involucral glands without horns · · · · · · · · · · · · · · · 39. *usambarica*
7. Branches and leaves completely glabrous; capsules large c. 5.5 × 6.5 mm, exserted on a pedicel to 15 mm long · 35. *cyparissioides*
 - Branches minutely pubescent at least at the leaf-bases; capsules to 3.5 × 4 mm, exserted on a pedicel c. 5 mm long · 8
8. Leaves oblanceolate up to 40 × 5 mm · 32. *whyteana*
 - Leaves linear, 2 × 1 mm to 20 × 3 mm · 9
9. Leaves up to 20 mm long; floral bracts rhomboid and apiculate, up to 6 × 6 mm · · · · · · ·
 · 33. *epicyparissias*
 - Leaves up to 10 mm long; floral bracts ovate, not apiculate, up to 4 × 3 mm · · 34. *crebrifolia*
10. Herbs with simple or sparsely-branched stems 30–60(100) cm long, from a woody rootstock
 · 37 *depauperata*
 - Woody shrubs to 3 m high · 38. *ugandensis*

29. **Euphorbia peplus** L., Sp. Pl.: 456 (1753). —N.E. Brown in F.C. **5**, 2: 255 (1915). —S. Carter in F.T.E.A., Euphorbiaceae, part 2: 432 (1988). Syntypes from Europe.

Annual herb to 25(35) cm high, glabrous. Leaves with a petiole to 1 cm long, shorter above; lamina to 25 × 15 mm, obovate, rounded at the apex, cuneate at the base, entire at the margin. Cymes axillary and in terminal 3-branched umbels forking many times, with primary rays to 3.5 cm long. Bracts sessile, similar to the leaves but more ovate. Cyathia on peduncles to 1.3 mm long, 1 × 1 mm with cup-shaped involucres; glands 4, 0.5 mm wide, transversely oblong with 2 horns to 0.8 mm long; lobes rounded, minute, margins ciliate. Male flowers: bracteoles linear, with minutely ciliate tips; stamens 1 mm long. Female flower: ovary glabrous; styles 0.3 mm long, joined at the base, spreading with deeply bifid apices. Capsule exserted on a pedicel to 3 mm long, 2 × 2 mm, deeply 3-lobed with truncate base and with longitudinal fleshy ridges each side of the sutures. Seeds 1.5 × 1 mm, oblong-ovoid with longitudinal rows of deep pits, reddish-brown becoming grey; caruncle 0.3 mm across, smooth, white.

Zambia. W: Kitwe, Coral Nurseries, fl. & fr. 8.iii.1967, *Anton-Smith* in *GHS* 184721 (SRGH).
Zimbabwe. W: Bulawayo Municipal Park, fl. & fr. 30.iv.1958, *Drummond* 5524 (EA; K; PRE; SRGH). C: Gweru, Harben Park, fl. & fr. 26.ii.1978, *Cannell* 735 (K; PRE; SRGH).
An introduced weed of cultivation; 1220–1400 m.

30. **Euphorbia schimperiana** Scheele in Linnaea **17**: 344 (1843). —Boissier in de Candolle, Prodr. **15**, 2: 155 (1862). —N.E. Brown in F.T.A. **6**, 1: 533 (1911). —S. Moore in J. Linn. Soc., Bot. **40**: 189 (1911). —Eyles in Trans. Roy. Soc. South Africa **5**: 400 (1916). —Robyns, Fl. Sperm. Parc Nat. Alb. **1**: 482 (1948). —Andrews, Fl. Pl. Sudan **2**: 75 (1952). —Brenan in Mem. New York Bot. Gard. **9**: 66 (1954). —Keay in F.W.T.A., ed. 2, **1**: 421 (1958). — Brummitt in Wye Coll. Malawi Proj. Rep.: 56 (1973). —Agnew, Upland Kenya Wild Fl.: 222 (1974). —S. Carter in Kew Bull. **40**: 812 (1985); in F.T.E.A., Euphorbiaceae, part 2: 433 (1988). Type from N Yemen.

Much-branched annual or short-lived perennial herb, erect to 2 m high, completely glabrous, or pubescent with long crisped hairs at least on the stem below the leaves. Leaves sessile, usually crowded and leaving prominent scars; lamina to 15 × 2 cm, ovate-lanceolate to lanceolate, apiculate at the apex, cuneate at the base, entire at the margin, glabrous or sparsely hairy on the lower surface. Cymes axillary and in terminal 3–15-branched umbels forking many times with primary rays to 15 cm long. Bracts sessile, 1–4 × 1–2 cm, deltoid or occasionally 3-lobed, apiculate, but the bracts below the umbel similar to the leaves. Cyathia on peduncles 0.5–3 mm long, 2 × 2 mm with cup-shaped involucres, glabrous or occasionally sparsely hairy; glands 4, spreading, 1 × 1.5–2 mm, transversely elliptic to reniform, 2-horned, the horns 0.5–1.5 mm long, green becoming brownish-red; lobes 0.5 mm long, subquadrate, shallowly 2-toothed at the apex, ciliate at the margin. Male flowers: bracteoles linear, with ciliate apices; stamens to 4.5 mm long, glabrous or occasionally hairy below the articulation. Female flower: ovary glabrous or pilose; styles erect to 2.5 mm long, joined at the base for 0.5 mm, bifid for up to 1 mm with apices spreading. Capsule exserted on a reflexed pedicel up to 5.5 mm long, up to 4 × 4.5 mm, deeply 3-lobed with truncate base, glabrous or sparsely pilose. Seeds 2–2.5 × 1.5–2 mm, oblong, slightly compressed, smooth, shiny black becoming grey; caruncle 0.5 mm across, wrinkled, yellowish.

1. Stems glabrous except sometimes for a few hairs in the leaf axils · · · · i) var. *schimperiana*
– Stems pubescent at least below the insertion of the leaves on the stem · · · · · · · · · · · · · 2
2. Capsule glabrous · ii) var. *pubescens*
– Capsule pilose · iii) var. *velutina*

i) Var. **schimperiana** —S. Carter in F.T.E.A., Euphorbiaceae, part 2: 433, fig. 80 (1988).
    Euphorbia dilatata A. Rich., Tent. Fl. Abyss. **2**: 240 (1851). Type from Ethiopia.
    Euphorbia monticola A. Rich., Tent. Fl. Abyss. **2**: 242 (1851). Type from Ethiopia.
    Euphorbia ampla Hook.f. in J. Linn. Soc., Bot. **6**: 20 (1861). —N.E. Brown in F.T.A. **6**, 1: 532 (1911). Type from Equatorial Guinea.
    Euphorbia ampla var. *tenuior* Hook.f., in J. Linn. Soc., Bot. **7**: 215 (1864). Syntypes from Cameroon.
    Euphorbia longecornuta Pax in Engler, Hochgebirgsfl. Afrika: 286 (1892). —N.E. Brown in F.T.A. **6**, 1: 535 (1911). —Robyns, Fl. Sperm. Parc Nat. Alb. **1**: 481 (1948). Type from Ethiopia.
    Euphorbia kilimandscharica Pax in Engler, Hochgebirgsfl. Afrika: 287 (1892). Type from Tanzania.

Euphorbia hochstetteriana Pax in Bot. Jahrb. Syst. **19**: 123 (1894), nom. illegit. Type as for *E. schimperiana*.
Euphorbia preussii Pax in Bot. Jahrb. Syst. **19**: 123 (1894). Type from Equatorial Guinea.
Euphorbia stuhlmannii Pax in Bot. Jahrb. Syst. **23**: 535 (1897). Type from Tanzania.
Euphorbia lehmbachii Pax in Bot. Jahrb. Syst. **28**: 27 (1899). Type from Equatorial Guinea.
Euphorbia helioscopia sensu Eyles in Trans. Roy. Soc. South Africa **5**: 399 (1916), non L.

Plants glabrous, or occasionally the stems of young shoots sparsely pilose and a few hairs present in the axils of the leaves.

Zambia. E: Chama Distr., Nyika Plateau, fl. & fr. 13.viii.1975, *Pawek* 10024 (K; MAL; MO). **Zimbabwe**. W: Shangani, fl. & fr. iii.1943, *Feiertag* in *GHS* 45458 (SRGH). E: Chimanimani Distr., Tarka Forest Reserve, fl. & fr. 27.xi.1967, *Simon & Ngoni* 1361 (K; PRE; SRGH). **Malawi**. N: Rumphi Distr., Nyika Plateau, Kafwimbi (Kafwimba) Forest, fl. & fr. 24.viii.1977, *Pawek* 12961 (K; MAL; MO). C: Dedza Distr., Chongoni Forest Reserve, fl. & fr. 27.iii.1968, *Salubeni* 1027 (K; PRE; SRGH). S: Blantyre, fl. & fr. 6.vii.1879, *Buchanan* 162 (K).
A common weed throughout tropical Africa. In montane and submontane grassland, forest edges and clearings; 760–2180 m.

ii) Var. **pubescens** (N.E. Br.) S. Carter in Kew Bull. **40**: 813 (1985); in F.T.E.A., Euphorbiaceae, part 2: 435, fig. 80/5 (1988). Type from Uganda.
   Euphorbia longipetiolata Pax & K. Hoffm. in Bot. Jahrb. Syst. **45**: 241 (1910). Type from Cameroon.
   Euphorbia longecornuta var. *pubescens* N.E. Br. in F.T.A. **6**, 1: 535 (1911). —Robyns, Fl. Sperm. Parc Nat. Alb. **1**: 482 (1948).

As for var. *schimperiana*, but stems usually pubescent with long crisped hairs, sometimes present only below the insertion of the leaves and bracts on the stem. Lower surface of young leaves often sparsely pilose. Involucre glabrous or often pilose. Ovary and capsule always glabrous.

Zambia. E: Chama Distr., Nyika, fl. & fr. 24.xii.1962, *Fanshawe* 7217 (K; NDO; SRGH). **Zimbabwe**. E: Mutasa Distr., 1 km east of Mapokana Hills, fl. & fr. 16.xi.1980, *Pope & Müller* 1789 (SRGH). **Malawi**. N: Chitipa Distr., Nyika, eastern foot of Nganda, fl. & fr. 29.vii.1972, *Brummitt, Munthali & Synge WC* 84 (K; MAL; SRGH). S: Blantyre Distr., southwest side of Ndirande Mt., fl. & fr. 31.v.1970, *Brummitt* 11188 (K). **Mozambique**. Z: Milange, fl. & fr. 11.ix.1949, *Barbosa & Carvalho* 4038 (K). T: Angónia Distr., Monte Dómuè, fl. & fr. 8.vi.1980, *Stefanesco & Nyongani* 546 (SRGH).
Also in Nigeria, Cameroon, southern Sudan, East Africa and Uganda. Montane grassland and forest edges and clearings; 650–2285 m.

iii) Var. **velutina** N.E. Br. in F.T.A. **6**, 1: 534 (1911). —S. Carter in Kew Bull. **40**: 813 (1985); in F.T.E.A., Euphorbiaceae, part 2: 435, fig. 80/6 (1988). Syntypes from Tanzania.
   Euphorbia velutina Pax in Engler, Pflanzenw. Ost-Afrikas **C**: 242 (1895), nom. illegit. Type from Tanzania.
   Euphorbia buchananii Pax in Bot. Jahrb. Syst. **28**: 27 (1899). Type: Malawi, Blantyre, *Buchanan* 7058 (PRE, holotype).
   Euphorbia schimperiana var. *buchananii* (Pax) N.E. Br. in F.T.A. **6**, 1: 534 (1911). —S. Moore in J. Linn. Soc., Bot. **40**: 189 (1911).

As for var. *schimperiana*, but all parts of the plant, at least on young growth, pilose with long crisped brownish hairs, usually very sparsely so on older growth and the main stem glabrescent. Involucre glabrous or pilose. Ovary and capsule always pilose.

Zimbabwe. E: Chirinda, fl. & fr. 23.x.1947, *Sturgeon* in *GHS* 18100 (K; SRGH). **Malawi**. S: Chikwawa (Chibisa) to Tshinsunze (Tshinmuze), fl. & fr. ix.1859, *Kirk* s.n. (K).
Also in East Africa. Evergreen rainforest edges, often near water; 1200–1760 m.
A widespread species in which all combinations of variable characters can be found throughout its distribution. The size of the plant and of its leaves, bracts and cyathia varies considerably depending upon local conditions of shade and moisture. Cyathial gland size and the length of the horns can vary on an individual plant, as also can the size of capsule and seeds. Whether the staminal pedicels are glabrous or pubescent and the length of the capsular pedicels likewise hold no significance. All these variations have been used in the past to distinguish distinct species which are placed here in synonymy. Pubescence of the stem, especially below the insertion of the leaves, and of the capsule, appear to be somewhat more constant characters and are thus used to differentiate the above varieties. However, in the Flora

Zambesiaca area these distinguishing features are not as marked as amongst most plants in East Africa, especially in var. *velutina* and the varietal limits are consequently less clear-cut.

31. **Euphorbia citrina** S. Carter in Kew Bull. **45**: 331 (1990). TAB. **75**, fig. A. Type: Zimbabwe, Nyanga, foot of Nyangani (Inyangani) Mt. near source of Matenderere (Matendere) R., 2220 m, fl. & fr. 1.viii.1988, *Carter & Coates-Palgrave* 2078 (K, holotype; EA; SRGH).
  Euphorbia sp. of Wild in Kirkia **1**: 58 (1960).

Straggly shrub 0.5–3 m high; branches minutely pubescent, leafless below and marked with crowded leaf scars. Leaves subsessile, glabrous except sometimes for a few hairs at the base; petiole c. 1 mm long, flattened, pubescent; lamina to 4.5 × 1 cm, lanceolate to linear-lanceolate, rounded at the apex and minutely apiculate, cuneate at the base, entire at the margin. Cymes axillary and terminal, 5–15 clustered in pseudumbels at the branch apices, each 2–4-forked, with primary rays to 3 cm long. Bracts sessile, c. 10 × 8 mm, ovate, yellow. Cyathia on peduncles 1–2 mm long, c. 3.5 × 4.5 mm, with cup-shaped involucres; glands 4, spreading, c. 1.3 × 2.5 mm, suborbicular or crescent-shaped with 2 blunt points or 2 horns to 0.8 mm long; lobes erect, 1.3 × 1.3 mm, quadrangular, apex distinctly 2-toothed with ciliate margins. Male flowers: bracteoles linear, minutely pubescent; stamens 5 mm long. Female flower: ovary glabrous; styles 2.5 mm long, joined for 1 mm, with spreading slightly thickened bifid apices. Capsule exserted on a reflexed pedicel to 7 mm long, 4.5 × 5 mm, deeply 3-lobed. Seeds 2.8 × 2 mm, oblong, smooth, shiny black; caruncle 1 mm across.

Zimbabwe. E: Nyanga Distr., Mtarazi Falls, fl. 21.vii.1947, *Wild* 1971 (K; SRGH). **Mozambique**. MS: Tsetserra, fl. 6.vi.1971, *Biegel & Pope* 3563 (SRGH).
  Confined to the Zimbabwe/Mozambique border mountains, between Nyanga and Chimanimani. In montane grassland usually at margins of evergreen forest; 1530–2100 m.

32. **Euphorbia whyteana** Baker f. in Trans. Linn. Soc. London, Bot. **4**: 39 (1894). —N.E. Brown in F.T.A. **6**, 1: 540 (1911). Type: Malawi, Mt. Mulanje, 1830 m, fl. 1891, *Whyte* 24 (K, holotype; BM).

Straggly perennial herb from a woody rootstock. Stems numerous, sparsely branching erect, to 1 m high. Branches minutely velvety-puberulous, rarely glabrous. Leaves crowded, sessile, glabrous; lamina to 40 × 5 mm, oblanceolate, obtuse at the apex and minutely apiculate, cuneate at the base, entire at the margin, midrib prominent beneath. Cymes axillary and terminal, 5–10 clustered in pseudumbels at the branch apices, each 2–3-forked, with primary rays to 2 cm long. Bracts sessile, c. 8 × 5 mm, ovate to rhomboid, yellow. Cyathia on peduncles c. 1 mm long, 2.5 × 3.5 mm, with cup-shaped involucres; glands 4(5), spreading, c. 1 × 1.5 mm, transversely elliptic, margin shallowly crenulate to crescent-shaped; lobes erect, 1.3 mm long, oblong with 2 rounded teeth at the apex, margins ciliolate. Male flowers: bracteoles linear, minutely pubescent; stamens 3.5 mm long, with pedicels pubescent. Female flower: ovary glabrous; styles 2 mm long, joined at the base, with spreading bifid apices. Capsule not seen.

Malawi. S: Mt. Mulanje, between Thuchila (Tuchila) Cottage and Ruo Valley Divide, fl. 14.vii.1956, *Jackson* 1918 (K); Mulanje Mt., fl. 23.vi.1958, *Chapman* H/686 (K; SRGH).
  Confined to Mulanje Mt. Locally common in montane tussock grassland; 1830–2290 m.

33. **Euphorbia epicyparissias** (Klotzsch & Garcke) Boiss. in de Candolle, Prodr. **15**, 2: 168 (1862). —N.E. Brown in F.C. **5**, 2: 266 (1915). Type from South Africa (Cape Province).
  Tithymalus epicyparissias Klotzsch & Garcke in Klotzsch, Nat. Pflanzenk. Tric.: 88 (1860).
  Tithymalus involucratus Klotzsch & Garcke in Klotzsch, Nat. Pflanzenk. Tric.: 91 (1860). Syntypes from South Africa (Cape Province).
  Euphorbia involucrata (Klotzsch & Garcke) Boiss. in de Candolle, Prodr. **15**, 2: 168 (1862).
  Euphorbia wahlbergii Boiss. in de Candolle, Prodr. **15**, 2: 169 (1862) nom. illegit. Type as for *Euphorbia epicyparissias*.
  Euphorbia epicyparissias var. *wahlbergii* N.E. Br. in F.C. **5**, 2: 266 (1915) nom. inval. Type as for *Euphorbia epicyparissias*.

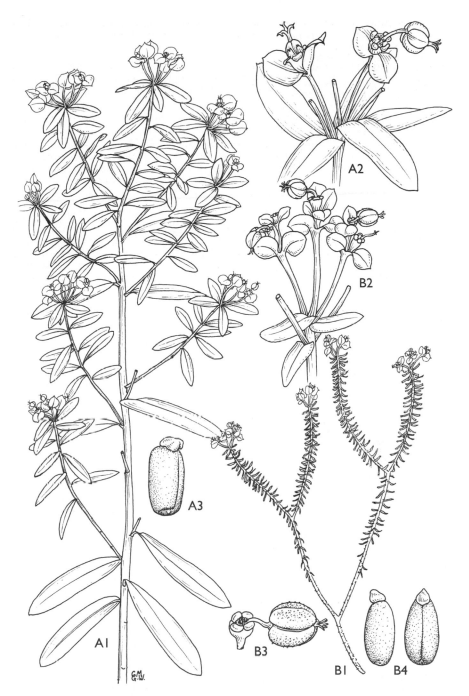

Tab. 75. A. —EUPHORBIA CITRINA. A1, fertile branch (× ⅔); A2, detail of pseudumbel (× 2); A3, seed (× 6), A1–A3 from *Methuen* 31. B. —EUPHORBIA CREBRIFOLIA. B1, fertile branch (× ⅔); B2, detail of pseudumbel (× 2), B1 & B2 from *Whellan & Davies* 968; B3, cyathium with capsule (× 4); B4, seeds (× 9), B3 & B4 from *Fries et al.* 5016. Drawn by Christine Grey-Wilson. From Kew Bull.

A many-branched shrubby perennial herb, 0.2–1 m high, from a woody shortly rhizomatous rootstock; branches glabrous or usually minutely pubescent towards the apex. Leaves crowded, subsessile, spreading or reflexed, glabrous except for a minute pubescence usually apparent at the base; lamina 10–20 × 1–3 mm, linear, apiculate at the apex, abruptly cuneate at the base, with thickened and ± revolute margins. Cymes terminal in 3–5-branched umbels, with primary rays to 1 cm long and usually minutely pubescent, each 1–3-forked, with 1 or more branches often developing as leafy shoots. Bracts to c. 6 × 6 mm, rhomboid, apiculate. Cyathia subsessile, c. 2.5 × 4.5 mm, with cup-shaped involucres; glands 4, c. 1 × 2 mm, transversely oblong, margin crenulate, crescent-shaped or shortly 2-horned; lobes 0.8 mm long, rounded, margin deeply 2-toothed, finely ciliate. Male flowers: bracteoles linear, ciliate; stamens 3.5 mm long. Female flower: ovary glabrous; styles 1.5 mm long, joined at the base, with spreading obviously bifid apices. Capsule exserted on a pedicel to 6 mm long, c. 3.5 × 4 mm, deeply 3-lobed. Seeds c. 2.8 × 1.8 mm, ovoid, smooth, black; caruncle 0.5 mm across.

Mozambique. GI: Bilene Distr., between Macia and south S. Martinho do Bilene, fl. & fr. 11.ii.1959, *Barbosa & Lemos* 8364 (COI; K; SRGH). M: Matutuíne Distr., Matutuíne (Bela Vista), Zitundo and Ponta do Ouro, fl. & fr. 25.vi.1968, *Balsinhas* 1301 (COI).
Apparently confined to the southern tip of Mozambique, but common in South Africa, especially KwaZulu-Natal and eastern Cape. In grassland, especially alongside rivers and streams; 50–150 m.
Measurements for this description were taken from the few specimens seen from southern Mozambique. In South Africa, where it is common and widespread, measurements vary considerably. Plants occurring in sandy soil away from damp areas, are shorter, more compact and with smaller, more crowded leaves. Conversely, leaves and especially the involucral bracts can be much larger when plants occur in damp shady situations. The pubescence on the branches of South African plants is likewise extremely variable, but appears to bear no relationship to ecological factors.

34. **Euphorbia crebrifolia** S. Carter in Kew Bull. **45**: 334 (1990). TAB. **75**, fig. B. Type: Zimbabwe, Nyanga Distr., foot of Nyangani (Inyangani) Mt. near source of Matenderere (Matendere) R., 2220 m, fl. & fr. 1.viii.1988, *Carter & Coates-Palgrave* 2077 (K, holotype; SRGH).
    Euphorbia ericoides sensu Wild in Kirkia **1**: 58 (1960), non Pax.

A sparsely-branched straggling herb, to c. 1 m high; branches slender, woody, minutely pubescent. Leaves subsessile; petiole flattened, to c. 1 mm long, pubescent; lamina reflexed, 2–10 × 1–2 mm, linear-lanceolate, minutely apiculate at the apex, rounded at the base, with margins entire and revolute. Cymes axillary and terminal, 4–8 clustered in pseudumbels at the branch apices, with primary rays to 1 cm long, each 1–2-forked, pubescent. Bracts sessile, to c. 4 × 3 mm, ovate, yellow. Cyathia c. 2 × 3 mm, with cup-shaped involucres; glands 4, spreading, c. 1 × 1.5 mm, transversely elliptic to ± crescent-shaped or occasionally with 2 minute horns; lobes 0.5 mm long, quadrangular, apex minutely 2-toothed, margin ciliolate. Male flowers: bracteoles laciniate, pubescent; stamens 2.3 mm long. Female flower: ovary glabrous; styles 1.2 mm long, joined at the base, with spreading bifid apices. Capsule exserted on a reflexed pedicel to 4.5 mm long, c. 3 × 3 mm, deeply 3-lobed. Seeds c. 2 × 1.3 mm, ovoid, smooth, shiny black; caruncle 0.6 mm across.

Zimbabwe. E: Chimanimani Mts., between Higher Valley and Binga (Point 71), fl. & fr. 25.ix.1966, *Simon* 823 (K; SRGH). **Mozambique**. MS: Chimanimani Mts., fl. & fr. 6.vi.1949, *Wild* 2904 (K; SRGH).
Known only from the Zimbabwe/Mozambique border mountains. Montane grassland amongst ericoid scrub, on rocky slopes; 1220–2285 m.

35. **Euphorbia cyparissioides** Pax in Bot. Jahrb. Syst. **19**: 123 (1894). —Hiern, Cat. Afr. Pl. Welw. **1**: 951 (1900). —N.E. Brown in F.T.A. **6**, 1: 542 (1911). —S. Moore in J. Linn. Soc., Bot. **40**: 189 (1911). —Eyles in Trans. Roy. Soc. South Africa **5**: 399 (1916). —Suessenguth & Merxmüller, [Contrib. Fl. Marandellas Distr.] Proc. & Trans. Rhod. Sci. Ass. **43**: 84 (1951). —Agnew, Upland Kenya Wild Fl.: 223 (1974). —S. Carter in Kew Bull. **40**: 814 (1985); in F.T.E.A., Euphorbiaceae, part 2: 439, fig. 81/1–3 (1988). Syntypes from Sudan.
    Euphorbia huillensis Pax in Bot. Jahrb. Syst. **28**: 27 (1899). Type from Angola.
    Euphorbia genistoides sensu Hiern, Cat. Afr. Pl. Welw. **1**: 952 (1900), non Berg.

Euphorbia ericifolia Pax in Bot. Jahrb. Syst. **33**: 288 (1903). Type from Tanzania.
Euphorbia dejecta N.E. Br. in F.T.A. **6**, 1: 541 (1911). Type: Malawi, Chitipa (Fort Hill), 1065–1220 m, fl. & fr. vii.1896, *Whyte* s.n. (K, holotype).
Euphorbia cyparissioides var. *minor* N.E. Br. in F.T.A. **6**, 1: 542 (1911). —Eyles in Trans. Roy. Soc. South Africa **5**: 399 (1916). —Suessenguth & Merxmüller, [Contrib. Fl. Marandellas Distr.] Proc. & Trans. Rhod. Sci. Ass. **43**: 84 (1951). Syntypes: Zambia, between Kabwe and Bwana Mkubwa Mine, fl. x.1906, *Allen* 354 (K; SRGH); fl. 24.xii.1907, *Kassner* 2125 (K). Zimbabwe, Chimanimani, *Swynnerton* 6043 (BM), 6044; Haroni R., *Swynnerton* 1529 (BM). Malawi, between Songwe and Karonga, fl. & fr. vii.1896, *Whyte* s.n. (K). Mozambique, Nhamazi (Inhamadzi) Valley, *Swynnerton* 1533 (BM). Others from Angola.
Euphorbia whyteana sensu Eyles in Trans. Roy. Soc. South Africa **5**: 400 (1916) as to *Swynnerton* 1526.

Glabrous perennial herb; rootstock woody, 1–2 cm thick, producing densely-tufted simple or sparsely-branched annual stems; stems erect, to 30 cm high, seldom more. Leaves densely crowded on the upper part of sterile growths, sparser and often shorter on flowering stems, sessile, spreading to reflexed; lamina to 30 × 5 mm, linear-lanceolate, rounded at the apex and conspicuously apiculate, cuneate at the base. Cymes occasionally axillary, usually clustered in terminal 3–7-branched pseudumbels with primary rays to 2 cm long, each forking usually once only or sometimes branching to produce further umbels. Bracts similar to the leaves, to 12 × 6 mm, and more ovate-lanceolate. Cyathia usually subsessile, but sometimes on peduncles to 5 mm long, especially those at the centre of the umbels which sometimes develop only male flowers, c. 3 × 5 mm with broadly cup-shaped involucres, the central cyathia larger to 8 mm in diameter; glands 4–7 or occasionally more and sometimes proliferating amongst the male flowers to produce a "double" involucre, 1–2 × 1.5–3 mm, transversely oblong, margin very irregular, rounded, shortly 2-horned, or crenulate and convoluted, bright yellow; lobes 1–1.5 mm long, rounded, margin shallowly toothed, finely ciliate. Male flowers very many: bracteoles linear, ciliate at the apex; stamens 3.5–4.5 mm long. Female flower: styles 1.8 mm long, joined for one-third, with spreading thickened bifid apices. Capsule exserted on a pedicel to 1.5 cm long, 5.5 × 6.5 mm, deeply 3-lobed, surface roughened. Seeds 3.5 × 3 mm, ovoid and slightly compressed, smooth, blackish with faint brown speckles; caruncle 1.5 mm across, wrinkled, orange.

Zambia. N: Mbala Distr., old Kasama road, fl. & fr. 27.vii.1960, *Richards* 12886 (K; SRGH). W: Mwinilunga Distr., Kabompo Gorge, fl. & fr. 22.xi.1962, *Richards* 17464 (K; SRGH). C: Serenje Distr., Kundalila Falls, fl. & fr. viii.1968, *G. Williamson* 1107 (SRGH). **Zimbabwe**. C: Dombi Estate, 20 km north of Marondera (Marandellas), fl. 13.xii.1970, *Biegel* 3432 (SRGH). E: Chimanimani Distr., Rocklands, fl. & fr. 7.x.1950, *Sturgeon & Panton* in *GHS* 30816 (K; SRGH). **Malawi**. N: Rumphi Distr., Nyika Plateau, 13 km from Chelinda Camp on road to main gate, fl. & fr. 7.vii.1970, *Brummitt* 11870 (K; SRGH). C: Kasungu National Park, fl. & fr. 23.xii.1970, *Hall-Martin* 1359 (SRGH). S: Mulanje Reserve, Likungusi, st. 14.viii.1930, *Topham* 234 (MAL). **Mozambique**. MS: Manica Distr., Gorongo Mt., Nhamucarara (Nyamakwarara) Valley, fl. & fr. 2.xi.1967, *Mavi* 440A (K; LISC; SRGH).
Widespread, also in Cameroon, Sudan, Ethiopia and southwards to East Africa, Dem. Rep. Congo and Angola. It is especially common in southeast Tanzania, northern Zambia and northern Malawi. Grassland and open woodland on well-drained soils, usually appearing after burning; 1005–2285 m.
This is an extremely variable taxon, like most widespread species. The short stems which usually appear after burning can often persist and elongate. Leaves are mostly needle-like and 10–15 mm in length, but can be scarcely 5 mm long and strongly reflexed, or spreading and much larger and broader. The cyathium varies considerably in size, and the glands especially, vary in number, size and shape.

36. **Euphorbia daviesii** E.A. Bruce in Bull. Misc. Inform., Kew **1940**: 51 (1940). —S. Carter in F.T.E.A., Euphorbiaceae, part 2: 437 (1988). Type from Tanzania.
Euphorbia imbricata E.A. Bruce in Bull. Misc. Inform., Kew **1933**: 468 (1933), nom. illegit. non Vahl. Type as for *E. daviesii.*

Glabrous perennial herb; rootstock woody, c. 1 cm thick, producing numerous annual stems; stems usually unbranched, erect to 30 cm tall. Leaves sessile, slightly fleshy; lamina to 2.5 × 1 cm, oblanceolate, acute to ± obtuse at the apex, apiculate, with entire margins. Cymes occasionally axillary, usually in terminal 3–7-branched umbels, with primary rays to 3 cm long, each forking c. 4 times. Bracts sessile, yellowish-green, to 1.5 × 1.5 cm, deltoid, obtuse to rounded at the apex, apiculate, subcordate at the base. Cyathia sessile, 3 × 3.5 mm, with cup-shaped involucres; glands 4, rarely 5 on those primary cyathia which develop all-male flowers, spreading, 1.5 × 2–2.5 mm, transversely oblong, outer margin entire to shallowly sinuate with 2 minute horns, golden-yellow; lobes rounded, 1 mm long with 2 shallow teeth at the apex, ciliate at the margin. Male flowers many: bracteoles ligulate, with ciliate apices; stamens 4 mm long. Female flower: styles 1.5 mm long, joined at the base, with spreading apices bifid for one-third. Capsule exserted on a pedicel to 6 mm long, 4 × 4.5 mm, conspicuously 3-lobed. Seeds 3 × 2–4 mm, ovoid, smooth, grey, speckled with brown; caruncle 0.5 mm across, cream-coloured.

Malawi. N: Rumphi Distr., Nyika Plateau, Kasaramba, fl. & fr. 17.xi.1967, *Richards* 22584 (K; SRGH).

Also in southern Tanzania. Montane grassland, in dry sandy soils; 2215–2350 m.

37. **Euphorbia depauperata** A. Rich., Tent. Fl. Abyss. **2**: 241 (1851). —Boissier in de Candolle, Prodr. **15**, 2: 119 (1862). —N.E. Brown in F.T.A. **6**, 1: 537 (1911). —S. Moore in J. Linn. Soc., Bot. **40**: 189 (1911). —Eyles in Trans. Roy. Soc. South Africa **5**: 399 (1916). —Brenan in Mem. New York Bot. Gard. **9**: 66 (1954). —Keay in F.W.T.A., ed. 2, **1**: 421 (1958). — Agnew, Upland Kenya Wild Fl.: 223 (1974). —S. Carter in Kew Bull. **40**: 815 (1985); in F.T.E.A., Euphorbiaceae, part 2: 439 (1988). Syntypes from Ethiopia.

Perennial herb; rootstock thick woody, producing numerous simple or sparsely branched annual stems; stems erect to 60 cm high, or spreading and decumbent, 30–100 cm long, rarely more, glabrous or pilose with long spreading hairs. Leaves subsessile; petiole c. 1 mm long, flattened, with a short tuft of hairs in the axil; lamina to 5–8 × 1–3 cm, linear-lanceolate to obovate or broadly obovate, obtuse to rounded at the apex and usually shortly apiculate, abruptly cuneate to rounded at the base, narrowly cartilaginous and sometimes ± revolute at the margin, glabrous. Cymes axillary and in (3)5(6)-branched umbels around a terminal bisexual cyathium, with primary rays to 7(10) cm long, each forking c. 3 times. Bracts sessile, c. 1.5–2.5 × 1.5–3 cm, deltoid to suborbicular, longer below the umbel, apex obtuse to rounded, base with a tuft of hairs on the upper surface. Cyathia sessile, glabrous or occasionally sparsely pilose, c. 3.5 × 5.5 mm, with broadly cup-shaped involucres; glands 4–6, spreading, 1.5–3 mm broad, orbicular to transversely oblong, yellow; lobes 1 mm long, rounded, densely ciliate on the inner surface. Male flowers: bracteoles fan-shaped, densely ciliate; stamens 4.5 mm long. Female flower: ovary densely verrucose, glabrous; styles 3 mm long, joined to nearly halfway, with spreading deeply bifid apices. Capsule exserted on a pedicel to 6 mm long, c. 5 × 6.5 mm, shallowly 3-lobed, densely and strongly verrucose, usually tinged reddish. Seeds c. 2.5 × 2 mm, ovoid, slightly compressed, smooth, greyish-brown; caruncle 1 mm across.

1. Apices of sterile shoots pilose · ii) var. *trachycarpa*
–   Apices of sterile shoots glabrous · 2
2. Leaves to 1(1.5) cm wide, linear-lanceolate to obovate, apices obtuse, apiculate · · · · · · · ·
· i) var. *depauperata*
–   Leaves to 3 cm wide, broadly obovate, apices rounded, not apiculate · · iii) var. *tsetserrensis*

i) Var. **depauperata** —S. Carter in F.T.E.A., Euphorbiaceae, part 2: 440, fig. 81/4–6 (1988).
      Euphorbia shirensis Baker f. in Trans. Linn. Soc., London, Bot. **4**: 38 (1894). Type: Malawi, Mt. Mulanje, fl. & fr., 1891, *Whyte* s.n. (K, holotype; BM).
      Euphorbia lepidocarpa Pax in Bot. Jahrb. Syst. **33**: 387 (1903). Type from Ethiopia.
      Euphorbia depauperata subsp. *aprica* Pax in Bot. Jahrb. Syst. **39**: 631 (1907). —N.E. Brown in F.T.A. **6**, 1: 538 (1911) [as var. *aprica*]. Type from Ethiopia.
      Euphorbia multiradiata Pax & K. Hoffm. in Bot. Jahrb. Syst. **45**: 240 (1910). Type from Cameroon.
      Euphorbia depauperata var. *pubiflora* N.E. Br. in F.T.A. **6**, 1: 538 (1911). Syntypes: Malawi, Nyika Plateau 1800–2100 m, fl. & fr. vi.1896, *Whyte* s.n. (K); vii.1896, *Whyte* s.n. (K); between Mpata and commencement of the Tanganyika Plateau, fl. & fr. vii.1896, *Whyte* s.n. (K).
      Euphorbia whyteana sensu S. Moore in J. Linn. Soc., Bot. **40**: 189 (1911).

Stems erect to c. 30 cm high, usually decumbent, glabrous or very sparsely pilose, with the apices of young sterile shoots glabrous. Leaves glabrous, to 1(1.5) cm wide, linear-lanceolate to obovate; apices obtuse, apiculate.   Bracts to 1.5 × 1.5 cm. Involucres glabrous or pilose.

Zambia. N: Mbala Distr., Lunzua Swamp, Chitembwa road, fl. & fr. 20.viii.1970, *Sanane* 1338 (K; SRGH). **Zimbabwe**. E: Chimanimani Mts., Outward Bound School, fl. & fr. 23.ix.1966, *Simon* 904 (K; PRE; SRGH). **Malawi**. N: Nyika Plateau, track to Kasenga Chipopoma Waterfall, fl. & fr. 13.xi.1967, *Richards* 22521 (K). C: Dedza Mt., fl. & fr. 16.viii.1976, *Pawek* 11593A (K; MAL; MO). S: Mulanje Mt., path to Lichenya (Luchenya) Plateau, fl. & fr. 5.vi.1962, *Richards* 16528 (K; SRGH).

Widespread throughout tropical Africa. Montane grassland in sandy rocky soil, and in forest clearings, also in watershed swampy grassland; 600–2195 m.

A pyrophytic species which exhibits a great deal of variation, particularly in stem-length, leaf-shape and pubescence. It is especially common in southern Tanzania and northern Malawi, where it often appears after burning, flowering when only a few centimetres high. The stem can persist and elongate, remain erect in tall grass, or become decumbent and creeping. The stems, leaves and involucres are usually completely glabrous in this variety, or one or all of these can bear scattered hairs. The stems of young sterile shoots are always completely glabrous.

ii) Var. **trachycarpa** (Pax) S. Carter in Kew Bull. **40**: 815 (1985); in F.T.E.A., Euphorbiaceae, part 2: 441, fig. 81/7 (1988). Type from Tanzania.
      Euphorbia trachycarpa Pax in Bot. Jahrb. Syst. **33**: 288 (1903).

As for var. *depauperata*, but stems and apices of young sterile shoots ± densely pilose, flowering stems usually more sparsely pilose. Leaves sometimes with long spreading hairs on the lower surface, especially along the midrib. Involucre glabrous or pilose.

Zambia. E: Chama Distr., Nyika Plateau, fl. & fr. 24.xi.1955, *Lees* 84 (K; NDO). **Malawi**. N: Rumphi Distr., Livingstonia, Nyamkhowa (Nyamkowa), fl. & fr. 23.ii.1978, *Pawek* 13854 (K; MAL; MO; PRE; SRGH).

Also in Tanzania, but is most common in northern Malawi.   In montane grassland; 1555–2135 m.

The stems of sterile leafy shoots are always pilose in this variety.

iii) Var. **tsetserrensis** S. Carter in Kew Bull. **45**: 336 (1990). Type: Zimbabwe, Mutare Distr., Himalaya Mts., Dicker's Farm, fl. & fr. 28.xi.1966, *Dale* SKF 433 (SRGH, holotype; K).

As for var. *depauperata*, but stems stouter and up c. 60 cm tall. Leaves subsessile, to 8 × 3 cm, broadly obovate; apex rounded, not apiculate; base rounded. Bracts of the inflorescence 1–2.5 × 1.5–3 cm. Involucre sparsely pilose.

Zimbabwe. E: Mutare Distr., Himalaya Mts., Tsetserra Farm, fl. & fr. 15.ii.1962, *Drewe* 32 (SRGH). **Mozambique**. MS: Sussundenga Distr., Tsetserra Mt., 2140 m, 7.ii.1955, *Exell, Mendonça & Wild* 223 (SRGH).

Apparently restricted to the Tsetserra and Himalaya Mts. in the Zuira Range, and reported to be locally common there. Montane grassland; 1500–2200 m.

Robust forms of the typical variety, with obovate leaves, have been collected on the Chimanimani Mts., but they are not as conspicuously large as the specimens above. The typical variety has not been collected from the Himalaya Mts.

38. **Euphorbia ugandensis** Pax in Bot. Jahrb. Syst. **45**: 240 (1910). —N.E. Brown in F.T.A. **6**, 1: 531 (1911). —Dale & Greenway, Kenya Trees & Shrubs: 202 (1961). —Agnew, Upland Kenya Wild Fl.: 223 (1974). —S. Carter in F.T.E.A., Euphorbiaceae, part 2: 441 (1988). Type from Kenya.

Shrubby perennial herb to 1(3) m high, woody at the base; branches in whorls of c. 5, glabrous or very sparsely pilose. Leaves subsessile with a very short flattened petiole; lamina to 11 × 3 cm, lanceolate, acute to shortly apiculate at the apex, cuneate at the base, entire at the margin, petiole and leaf lower surface pilose with long, soft hairs, especially the midrib and sometimes the upper surface of young leaves. Cymes in terminal 5-branched umbels, with primary rays to 7 cm long, each forking usually twice. Bracts sessile, to 2 × 2 cm, deltoid, apex obtuse to rounded and minutely apiculate, base and lower margins sparsely pilose. Central cyathium of the umbel on a peduncle 1.5 mm long, the rest sessile, 3 × 3.8 mm, with funnel-shaped involucres, glabrous; glands 4–5, c. 1.5 mm broad, transversely oblong and 2-lipped, yellow-green; lobes 1 mm long, subquadrate, apex with 2 shallow teeth, ciliate on the inner surface. Male flowers: bracteoles spathulate, deeply toothed, densely ciliate; stamens 4 mm long. Female flower: ovary glabrous, verrucose; styles 1.8 mm long, joined at the base, spreading, with deeply bifid thickened apices. Capsule exserted on a pedicel to 6 mm long, 4 × 5 mm, 3-lobed, almost smooth but slightly verrucose when mature, capsule wall very thick. Seeds 3.2 × 2.5 mm, ovoid, obscurely wrinkled, grey; caruncle 0.5 mm across, whitish.

Malawi. N: Chitipa Distr., Misuku Hills, Matipa Rainforest, fl. & fr. viii.1954, *Chapman* 312 (COI).
Also in southeast Uganda and southern Kenya, with a disjunct distribution in central southern Tanzania of which this Malawi collection represents the southernmost extension. Montane forest clearings, in damp situations; 1750–2050 m.
The measurements were taken mostly from East African specimens. The Malawi specimen is the only one seen from the Flora Zambesiaca area, from very close to the Tanzania border.

39. **Euphorbia usambarica** Pax in Bot. Jahrb. Syst. **19**: 122 (1894). —N.E. Brown in F.T.A. **6**, 1: 538 (1911). —Brenan, Check-list For. Trees Shrubs Tang. Terr.: 212 (1949). —Dale & Greenway, Kenya Trees & Shrubs: 202 (1961). —S. Carter in F.T.E.A., Euphorbiaceae, part 2: 441 (1988). Type from Tanzania.

Subsp. **usambarica**

Glabrous shrub to 3(4.5) m high, with long slender often subpendent branches. Leaves subsessile; lamina to 12 × 2 cm, oblanceolate, acute at the apex and markedly apiculate, cuneate at the base, entire, dark green and slightly fleshy. Cymes in terminal 3–5-branched umbels, with rays to 3 cm long, not forking. Bracts 6–10 mm long, suborbicular, apex obtuse to rounded, markedly apiculate. Central cyathium of the umbel on a peduncle to 2 mm long, the rest sessile to subsessile, 3 × 8 mm, with cup-shaped involucres; glands 4, or 5 on cyathia developing only male flowers (usually the central one), 1.5–2.5 mm broad, suborbicular to transversely elliptic, outer margin rounded and entire to truncate and shallowly crenulate; lobes 1.2 mm long, subquadrate with deeply 2-toothed apex, lower margins shortly ciliate. Male flowers: bracteoles fan-shaped, deeply divided, plumose; stamens 4 mm long. Female flower: styles 2.5 mm long, joined at the base, with spreading, distinctly bifid, slightly thickened apices. Capsule exserted on a pedicel 6–10 mm long, 6 × 8 mm, deeply 3-lobed. Seeds 3.4 × 3 mm, ovoid, lightly ridged and wrinkled, grey; caruncle 0.7 mm across.

Malawi. N: Nkhata Bay Distr., Lady Roseveare's house (Roseveare's), 8 km east of Mzuzu, fl. & fr. 12.viii.1972, *Pawek* 5635 (SRGH).
Also from southeast Kenya and eastern Tanzania and southwards to Lake Malawi; in the Flora Zambesiaca area known so far only from near Mzuzu. Open montane forest; c. 1280 m.
Subsp. *elliptica* Pax, with elliptical leaves, is known only from the western Usambara Mts., in Tanzania.

Sect. 8. TRICHADENIA Pax & K. Hoffm. in Engler, Pflanzenw. Afrikas [Veg. Erde 9] **3**, 2: 152 (1921).

Trees or shrubs, or perennial herbs with a fleshy rootstock and often fleshy stems. Leaves entire, sessile or petiolate; stipules apparently absent, or filamentous and soon deciduous, or modified as glands. Cyathia in terminal umbellate cymes; bracts sessile, deltoid, or leaf-like below the umbel. Involucres bisexual; glands 4 or 5, entire, or more usually crenulate, or 2-horned or with finger-like processes on the outer margin; lobes 5. Stamens clearly exserted from the involucre; bracteoles included, apices feathery. Perianth of female flower reduced to a rim below the ovary, or obvious and 3-lobed in those species with involucral glands producing marginal processes; styles with bifid apices. Capsule exserted, relatively large, deeply 3-lobed to subglobose. Seeds ovoid to subglobose, without a caruncle.

1. Involucral glands entire, crenulate or with 2 horns · 2
 – Involucral glands with 6–10 finger-like processes on the outer margin · · · · · · · · · · · · 10

2. Ovary and capsule glabrous · 3
   − Ovary and capsule pubescent or ± pilose · 7
3. Involucral glands with 2 long slender horns · · · · · · · · · · · · · · · · · · · 42. *dolichoceras*
   − Involucral glands entire, crenulate or lobed · 4
4. Bushy herbs or shrubs 0.5–3 m high, with semi-succulent branches · · · · · · · · · · · · · · · 5
   − Perennial herbs, with usually unbranched fleshy stems no more than 35 cm high · · · · 6
5. Leaves sparsely pilose on the lower surface and petiole · · · · · · · · · · · · · · · · · 40. *goetzei*
   − Leaves glabrous, or occasionally with a few hairs on the petiole and lower margins · · · · ·
   · 41. *transvaalensis*
6. Stems to 35 cm high bearing both leaves and flowers; seed surface with shallow rounded
   reticulations · 47. *platycephala*
   − Stems to 15 cm high, flowering separately from the leaves; seed surface with sharply ridged
   reticulations · 48. *eranthes*
7. Capsule densely pubescent-papillose · 8
   − Capsule thinly pilose or pubescent · 9
8. Floral bracts similar to the leaves, narrowly lanceolate; stems erect to 45 cm high · · · · · ·
   · 43. *ruficeps*
   − Floral bracts ovate and distinct from the ovate-lanceolate leaves on the flowering stems and
   the lanceolate leaves on the sterile stems; stems erect to 15 cm or decumbent to 35 cm · ·
   · 46. *papillosicapsa*
9. Cyme branches c. 3 mm long; bracts to 4 mm long, spathulate and distinct from the linear
   leaves · 44. *erythrocephala*
   − Cyme branches up to 8 cm long; bracts to 3 cm long, linear and leaf-like · · · · 45. *arrecta*
10. Perennial herb with annual stems to 12 cm high from a large tuber · · · · · 49. *trichadenia*
    − Shrub or small tree 1.5–9 m high with semi-succulent branches · · · · · · · · · · · 50. *grantii*

40. **Euphorbia goetzei** Pax in Bot. Jahrb. Syst. **28**: 420 (1900). —S. Carter in F.T.E.A., Euphorbiaceae, part 2: 454, fig. 84/5–8 (1988). Type from Tanzania.
    Euphorbia transvaalensis sensu N.E. Brown in F.T.A. **6**, 1: 530 (1911) pro parte as to "Afr. or." specimens, non Schlechter. —sensu F. White, F.F.N.R.: 199 (1962) pro parte as to *F. White* 3673, non Schlechter.

Shrubby perennial herb to 3(3.5) m high, with greyish semi-succulent stems and branches. Leaves with a pilose petiole to 4 cm long; stipules glandular; lamina to 17 × 6 cm, obovate, rounded at the apex, cuneate at the base, entire, glaucous especially beneath, sparsely pilose on the lower surface with long (1–1.5 mm) spreading hairs especially along the midrib, rarely completely glabrous in mature leaves. Cymes in 3–5-branched umbels on peduncles to 12 cm long, from whorls of 5–10 leaves at the swollen apices of branches, with primary rays to 8 cm long, each forking twice. Bracts c. 1.5 cm in diameter, the lower ones larger, subcircular, subsessile to shortly petiolate, glabrous except for a few hairs along the lower margins. Cyathia on peduncles to 1.5 mm long, 3 × 7 mm, with funnel-shaped involucres; glands 4, or 5 on the central cyathium of the umbel, spreading, c. 3 mm broad, transversely elliptic, the outer margin 2–4-lobed with each lobe sometimes shortly divided; involucral lobes c. 1 × 1.5 mm, broadly rounded, margin ciliate. Male flowers: bracteoles few, strap-shaped, upper margin ciliate and sometimes deeply divided; stamens 4.5–5 mm long. Female flower: styles 2.5–3.5 mm long, joined for 0.5–1 mm, suberect with spreading shortly bifid apices. Capsule exserted on a pedicel to 8 mm long, 8 × 10 mm, deeply 3-lobed with a sunken apex. Seeds c. 5.5 × 4 mm, ovoid with acute apex, surface with rather closely-set large flattened warts, brownish-grey.

Zambia. N: Mbala Distr., Lake Tanganyika, Crocodile Island, fl. 9.ii.1964, *Richards* 18995 (K). **Malawi**. N: Rumphi Gorge on South Rukuru R., c. 6.4 km east of Rumphi, fl. 12.ii.1968, *Simon, Williamson & Ball* 1787 (K; SRGH). C: 23 km southeast of Lilongwe by Ngala Mt., fr. 28.iv.1970, *Brummitt* 10237 (K). S: Machinga Distr., Liwonde National Park, fr. 17.iv.1980, *Blackmore, Brummitt & Banda* 1258 (K; MAL). **Mozambique**. N: 15 km from Pemba on road to Montepuez, fl. 27.i.1984, *Groenendijk, Maite & Dungo* 820 (K; MAL).
    Also in southern Ethiopia, Kenya and Tanzania; in Zambia it has been collected only in Mbala District. In thickets, usually by lake or river edges, amongst rocks; 450–1370 m.
    Since publication of the F.T.A. (**6**, 1: 530 (1911)), *E. goetzei* has been considered as synonymous with *E. transvaalensis*, which can however, be easily separated by its smaller size and glabrous leaves, except sometimes for a few hairs on the petiole.

41. **Euphorbia transvaalensis** Schltr. in J. Bot. **34**: 394 (1896). —N.E. Brown in F.T.A. **6**, 1: 530 (1911) pro parte excl. "Afr. or." specimens; in F.C. **5**, 2: 269 (1915). —White, Dyer & Sloane, Succ. Euphorb. **1**: 97 (1941). —F. White, F.F.N.R.: 199 (1962) pro parte excl. *F. White* 3673. Type from South Africa (Transvaal).

    Euphorbia galpinii Pax in Bull. Herb. Boissier **6**: 742 (1898), nom. illegit. Type as for *E. transvaalensis*.

    Euphorbia ciliolata Pax in Bull. Herb. Boissier **6**: 743 (1898). Type from Angola.

Shrubby perennial herb to 1.2 m high, with softly woody stems and branches. Leaves with a petiole to 2 cm long; stipules glandular; lamina to 11 × 5 cm, obovate, rounded at the apex, cuneate at the base, entire, glabrous, or occasionally with a few long spreading hairs on the petiole and lower margins. Cymes in 3(5)-branched umbels on peduncles to 15(19) cm long, from whorls of 4–7 leaves at the branch apices, with primary rays to 10 cm long, each forking usually once only. Bracts c. 1.5–2 mm in diameter, ovate to suborbicular, subsessile to shortly petiolate, glabrous. Cyathia on peduncles c. 1 mm long, 3 × 5 mm with funnel-shaped involucres; glands 4, or 5 on the central cyathium of the umbel, spreading, c. 2 mm broad, transversely elliptic, the outer margin 2-lobed; involucral lobes c. 1 × 1.5 mm, broadly rounded, margin ciliate. Male flowers: bracteoles few, strap-shaped, ciliate on upper margins and deeply divided; stamens 4 mm long. Female flower: styles 3.5–4.5 mm long, joined to halfway, erect with slightly spreading very shortly bifid apices. Capsule exserted on a pedicel to 6 mm long, 8 × 10 mm, deeply 3-lobed with a sunken apex. Seeds c. 5.5 × 4 mm, ovoid with an acute apex, surface sparsely covered with large flattened warts, brownish-grey.

    Caprivi Strip. Andara, fl. & fr. 17.i.1956, *de Winter* 4318 (K). **Botswana**. N: near Tshesebe (Tsessebe), fr. 8.iii.1965, *Wild & Drummond* 6814 (K; PRE; SRGH). **Zambia**. B: Sesheke Distr., 6 km south of Masese, fl. i.1973, *G. Williamson* 2273 (K; SRGH). C: Kabwe Distr., Mulungushi, fr. 9.ii.1964, *Fanshawe* 8258 (K; NDO). E: Chipata Distr., 9.6 km north of Msoro Mission, fl. 24.i.1930, *Bush* 15 (K). S: Gwembe Distr., Musaya R., 47 km west of Chirundu, fl. 9.xii.1971, *Kornaś* 622 (K). **Zimbabwe**. N: Mount Darwin Distr., Kandeya C.L. (Native Reserve), fl. 18.i.1960, *Phipps* 2324 (K; SRGH). W: Matobo Distr., Matopos Research Station, fl. & fr. 12.xii.1951, *Plowes* 1377 (K; SRGH). C: Harare Distr., Seke Dam (Prince Edward Dam), fl. & fr. 8.i.1952, *Wild* 3746 (K; SRGH). E: Chipinge Distr., 1 km west of confluence of Musirizwi and Bwazi Rivers, fr. 30.i.1975, *Biegel, Pope & Gibbs Russell* 4869 (K; SRGH). S: Bikita Distr., 8 km west of Moodie's Pass, fl. 28.i.1961, *Leach* 10715 (K; PRE; SRGH). **Mozambique**. T: Lupata, fl. & fr. i.1860, *Kirk* s.n. (K).

    Also in South Africa (Transvaal) and Namibia. Deciduous woodlands, in sandy soils often in rocky places and escarpments; 90–1400 m.

    E. transvaalensis and *E. goetzei* have long been confused, but the two species may be distinguished on morphological differences and geographical distribution. *E. goetzei* is a northern taxon extending only into the northern parts of the Flora Zambesiaca area, and is distinguished by its larger size and by its leaves sparsely pilose on the lower surface.

42. **Euphorbia dolichoceras** S. Carter in Kew Bull. **35**: 418 (1980); in F.T.E.A., Euphorbiaceae, part 2: 455 (1982). Type from Tanzania.

Glabrous perennial herb with a thick fleshy rhizomatous rootstock and annual stems to 60 cm high. Leaves sessile; stipules glandular, minute, rarely obvious; lamina to 9 × 1.5 cm, obovate-lanceolate, rounded at the apex and apiculate, cuneate at the base, entire. Cymes axillary and in 3–5-branched umbels, with primary rays to 12 cm long, each forking 3–4 times. Bracts sessile, c. 2 × 2 cm, the lower ones larger and longer, deltoid, apiculate at the apex, truncate to subcordate at the base. Cyathia on peduncles 2–3 mm long or the central peduncle of the umbel to 1.5 cm long, c. 3 × 5 mm, with cup-shaped involucres; glands 4, or 5 on the central cyathium, to 1.5 mm broad, transversely elliptic, with 2 horns 1.5–2 mm long; lobes c. 1 mm long, broadly deltoid, margin ciliate with long hairs. Male flowers: bracteoles deeply laciniate with feathery apices; stamens 4 mm long. Female flower: ovary pedicellate; styles 2 mm long, joined for 0.5 mm, with roughened shortly bifid apices. Capsule exserted on a pedicel to 1 cm long, 4 × 6 mm, shallowly 3-lobed. Seeds 3 × 2.5 mm, subglobose, with a longitudinal dorsal ridge and obscure rows of small sharply pointed warts.

Malawi. N: Mzimba Distr., 16 km east of Mzambazi, 25 km west of main road, fl. 30.xii.1975, *Pawek* 10669B (K). C: Mchinji Distr., Kachebere, fl. & fr. 7.i.1959, *Robson* 1064 (K; SRGH); Dowa Distr., Kongwe Hill Forest Reserve, fl. & fr. 7.iii.1982, *Brummitt, Polhill & Banda* 16379 (K).
Also in southeast Tanzania. On sandy loam in *Brachystegia* woodland; 1000–1500 m.

43. **Euphorbia ruficeps** S. Carter in Kew Bull. **35**: 416 (1980); in F.T.E.A., Euphorbiaceae, part 2: 455 (1988). Type: Zambia, Kitwe, fl. & fr. 7.xii.1955, *Fanshawe* 2639 (K, holotype; NDO).

Perennial herb, with a woody tuberous rootstock, producing a few shortly puberulous annual stems to 45 cm high. Leaves deflexed, with a puberulous petiole to 1 cm long; stipules glandular, purplish; lamina to 12 × 1.3 cm, narrowly lanceolate, obtuse at the apex and apiculate, rounded at the base, entire revolute at the margin, midrib prominent on the lower surface which is sparsely pubescent. Cymes in 3-branched umbels, with primary rays to 12 cm long, each forking twice. Bracts similar to the leaves. Cyathia sessile, 4 × 5.5 mm, with shortly puberulous purplish-red broadly cup-shaped involucres; glands 4, c. 4 mm wide, transversely elliptic, spreading, dark red; lobes 2 mm long, subquadrate with margins deeply toothed and ciliate. Male flowers: bracteoles laciniate with feathery apices; stamens 5.5 mm long, puberulous. Female flower: ovary densely pilose; styles 3 mm long, sparsely puberulous below, joined to halfway with bifid apices. Capsule exserted on a curved pedicel, 5 mm in diameter, subglobose, very densely papillose-pilose, purplish-red; pedicel densely pilose, 6 mm long. Seeds 4 × 2.5 mm, ovoid, closely warted, greyish-brown.

Zambia. N: Nchelenge Distr., Mweru–Tamba Pass, fl. 13.x.1949, *Bullock* 1249 (K). W: Kitwe, fr. 6.ii.1956, *Fanshawe* 2769 (K; NDO; SRGH). C: Mkushi Distr., Fiwila, fl. 8.i.1958, *E.A. Robinson* 2695 (K; SRGH). E: Lundazi Distr., Lukusuzi National Park, Mwezi R., fl. 2.xii.1970, *Sayer* 805 (SRGH). **Malawi**. N: Mzimba Distr., 11 km south of Euthini (Eutini), 8 km west towards Rukuru R., fl. & fr. 29.xii.1975, *Pawek* 10646 (K; MAL; SRGH). C: 40 km south of Kasungu, fl. 22.xii.1970, *Pawek* 4124 (K; MAL).
Also in western Tanzania. Sandy soil in *Brachystegia* woodland; 1035–1550 m.

44. **Euphorbia erythrocephala** P.R.O. Bally & Milne-Redh. in Hooker's Icon. Pl. **35**: t. 3480 (1950). Type: Zambia, Mwinilunga Distr., source of Matonchi Dambo, fl. 16.ii.1938, *Milne-Redhead* 4607 (K, holotype; BM; PRE).

Perennial herb, with numerous annual stems from a woody rootstock; rootstock vertical c. 13 mm thick, cylindrical; stems clustered, slender, erect to c. 1 m high, woody at the base, glabrous below, shortly pubescent at the apex and around the leaf bases. Leaves sessile; stipules glandular, purple, in dense clusters of minute knobs; lamina to 50 × 1.5 mm, linear, mucronate at the apex, cuneate at the base, revolute at the margin, very sparsely puberulous, more densely so at the base. Cymes terminal, consisting of a central cyathium and 1–3 lateral cyathia, with rays 3 mm long. Bracts sessile, to 4 × 2 mm, spathulate with emarginate and mucronate apices, puberulous, pink. Cyathia subsessile, c. 4 × 4 mm, with barrel-shaped involucres, pubescent; glands 5, erect, 2 mm wide, 2-lipped, the inner lip triangular, pubescent, the outer lip subquadrate, 2 mm long with upper margin recurved, crimson; lobes rounded, pubescent. Male flowers: bracteoles fan-shaped, deeply laciniate, feathery; stamens 5 mm long. Female flower: ovary puberulous; styles 2.5 mm long, pubescent at the base, joined for 2 mm, with spreading bifid apices. Capsule just exserted, not seen entire but probably c. 5 × 5 mm, thinly pubescent; pedicel pubescent, 4.5 mm long. Seeds 4.3 × 3.5 mm, ovoid, surface densely covered with tiny warts, especially prominent towards the apex, black at the base, yellowish at the apex.

Zambia. W: Mwinilunga, fr. 16.v.1969, *Mutimushi* 3172 (K; NDO; SRGH); source of Matonchi River, fl. 18.ii.1975, *Hooper & Townsend* 160 (K; NDO; SRGH).
Known only from the Matonchi area. Watershed plain and dambo on Kalahari Sand, sandy grassland and open woodland; c. 1300 m.

45. **Euphorbia arrecta** N.E. Br. in Fries, Wiss. Ergebn. Schwed. Rhod.-Kongo-Exped.: 116 (1914) [non F.C. **5**, 2: 283 (1915)]. —S. Carter in F.T.E.A., Euphorbiaceae, part 2: 456 (1988). Type: Zambia, Kuta, between Lakes Bangweulu and Tanganyika, fl. & fr. 22.x.1911, *R.E. Fries* 1078 (UPS, holotype; K, drawing of holotype).

Perennial herb, with annual stems from a branched cylindrical woody rootstock; stems up to 45 cm high, sparsely branched, puberulous. Leaves usually reflexed on a petiole to 8 mm long; stipules glandular, minute; lamina to 5 × 0.8 cm, linear-lanceolate, obtuse at the apex, rounded at the base, minutely serrate with glandular teeth on the margins, pilose with long spreading hairs, midrib prominent on the lower surface. Cymes in axillary and terminal 3-branched umbels, with primary rays to 8 cm long, each forking once. Bracts similar to the leaves. Cyathia sessile, 3.5 × 5 mm, with puberulous funnel-shaped involucres; glands 4–5, 2.5 mm wide, transversely elliptic, minutely crenulate on the margin, spreading, reddish; lobes 1 mm long, subquadrate with a fringed ciliate margin. Male flowers: bracteoles fan-shaped, deeply laciniate, feathery; stamens 4 mm long. Female flower: ovary densely pubescent; styles 2.5 mm long, joined to halfway, with spreading thickened apices. Capsule exserted on a pilose pedicel 3.5 mm long, 6.5 × 7.5 mm, obtusely 3-lobed with a truncate base, sparsely pilose. Seeds 4.8 × 3.5 mm, ovoid, acutely triangular at the apex, obscurely 4-angled, surface obscurely and closely wrinkled, purplish-black.

Zambia. N: Mbala Distr., Chilongowelo Farm, fl. & fr. 22.i.1955, *Richards* 4225 (K); 9.5 km from Mbala on Mbeya road, fl. 4.ix.1956, *Richards* 6107 (K).
Also in southwest Tanzania. *Brachystegia* woodland amongst grasses, often appearing after annual fires; 1150–1625 m.

46. **Euphorbia papillosicapsa** L.C. Leach in Bull. Jard. Bot. Belg. **45**: 206 (1975). Type: Zambia, Mwense Distr., 25 km north of Chipili Mission, fr. 27.ii.1970, *Drummond & Williamson* 10066 (SRGH, holotype; K; PRE).

Perennial herb with 1–2 woody underground stems from a tuberous rootstock; rootstock 3–4 cm in diameter, subspherical; stems c. 7 cm long, branching at ground level; branches several, rebranching only at the base, shortly erect to 15 cm high or decumbent to 35 cm long, pubescent. Leaves spreading to reflexed, with a pubescent petiole 1–2 mm long; stipules glandular, minute; lamina to 23 × 5 mm and lanceolate on sterile branches, or to 30 × 10 mm and ovate-lanceolate on flowering branches, acute at the apex, rounded at the base, very narrowly cartilaginous and serrulate at the margin, usually revolute, glabrous except for a few hairs at the base, midrib prominent beneath. Cymes in terminal 3–5-branched umbels with rays to 6 cm long, not forking, or cyathia occasionally solitary. Bracts 4 or 5, subsessile, to 25 × 15 mm, ovate, apex acute, base subcordate. Cyathia sessile, c. 4 × 7 mm, with funnel-shaped pubescent involucres; glands 5, spreading, c. 3.5 × 1.8 mm, transversely elliptic with obscurely crenulate margins; lobes 2 × 2 mm, rounded, finely toothed at the margin with the central tooth extended to 1 mm long, ciliate. Male flowers: bracteoles fan-shaped, deeply laciniate; stamens 5 mm long. Female flower: ovary densely pubescent; styles 2.5 mm long, joined to halfway, with spreading bifid apices. Capsule scarcely exserted on a pubescent pedicel c. 3 mm long, 7 × 8 mm, subspherical, densely papillose-pubescent, purplish-brown. Seeds 4 × 2.8 mm, ovoid-oblong, with obtuse apex, smooth, purplish-brown.

Zambia. N: Mbala, Sand Pits, fl. 4.i.1961, *Richards* 13772 (K; SRGH); Mbala Distr., Itimbwe (Itembwe) Gorge, fl. & fr. 3.i.1960, *Richards* 12064 (K; SRGH).
So far know only from the Chipili Mission and Mbala areas. *Brachystegia* woodland amongst grass, in sandy soil; c. 1200–1500 m.

47. **Euphorbia platycephala** Pax in Bot. Jahrb. Syst. **19**: 122 (1894). —N.E. Brown in F.T.A. **6**, 1: 525 (1911). —S. Carter in F.T.E.A., Euphorbiaceae, part 2: 454 (1988). TAB. **76**, fig. A. Type from Tanzania.
Euphorbia uehehensis Pax in Bot. Jahrb. Syst. **28**: 420 (1900). Type from Tanzania.

Glabrous perennial herb with annual stems from a fleshy rootstock; rootstock, 20–30 × 2–4 cm, cylindrical; stems subfleshy to 35 cm high, rarely persistent and

Tab. 76. A. —EUPHORBIA PLATYCEPHALA. A1, habit (× ²⁄₃); A2, cyathium (× 4), A1 & A2 from *Gassner & Williamson* 2315; A3, fruiting cyathium (× 1); A4, seed (× 4), A3 & A4 from *Brummitt* 8799. B. —EUPHORBIA WILDII. B1, flowering branch (× ²⁄₃); B2, cyathium (× 4), B1 & B2 from *Dyer* 13196. Drawn by Christine Grey-Wilson.

woody. Leaves sessile; stipules filamentous, c. 1 mm long, deciduous; lamina to c. 12 × 4 cm, lanceolate to obovate, rounded to obtuse at the apex and minutely apiculate, cuneate at the base, entire. Cymes axillary and in terminal 3-branched umbels with primary rays to 10 cm long, each forking 2–4 times. Bracts to 4.5 × 4 cm, deltoid, apex obtuse-apiculate, base subcordate, those at the base of the umbel similar to the leaves. Cyathia on peduncles to 1.5 mm long, 5.5 × 10 mm, with cup-shaped involucres; glands 5, spreading, 1.5 × 2–3.5 mm, transversely elliptic with outer margin obscurely crenulate; lobes 1.5 × 3 mm, rounded, margin entire and minutely ciliate. Male flowers many: bracteoles laciniate with feathery apices; stamens 8 mm long, with anther cells minutely hispid. Female flower: styles 2 mm long, joined at the base for 0.5 mm, then widely spreading, with shortly bifid apices. Capsule 1 × 1.2 cm, deeply 3-lobed; pedicel 2.5 mm long. Seeds c. 4.5 mm in diameter, globose; surface reticulately wrinkled, brownish-grey.

Zambia. B: Kabompo, fl. 22.ix.1964, *Fanshawe* 8919 (SRGH). W: Mwinilunga Distr., Kalenda Ridge, west of Matonchi Farm, fl. & fr. 8.x.1937, *Milne-Redhead* 2665 (K). E: Chipata Distr., Luangwa Valley, near Jumbe, fr. 21.i.1970, *Astle* 5756 (K; SRGH). S: Sikaonzwe, 71 km from Livingstone on Sesheke road, fl. 15.xii.1977, *Chisumpa* 420 (NDO). **Zimbabwe**. N: 19 km north of Gokwe, fr. 18.i.1964, *Bingham* 1090 (K; LISC; SRGH). W: Umguza Distr., Nyamandhlovu, fr. 15.ii.1956, *Plowes* 1925 (SRGH). **Malawi**. N: 35 km west of Karonga, fr. 26.iv.1977, *Pawek* 12703 (K). C: Salima Distr., north of Chitala on Kasache road, fr. 12.ii.1959, *Robson* 1575 (K; LISC; SRGH). S: Mangochi Distr., 6 km south of Monkey Bay, fr. 28.ii.1970, *Brummitt* 8799 (K; SRGH).
Also widely distributed in Tanzania. In seasonally wet areas, amongst rocks in sandy soils and hard clay; 480–1280 m.

48. **Euphorbia eranthes** R.A. Dyer & Milne-Redh. in Bull. Misc. Inform., Kew **1937**: 413 (1937).
Type: Zambia, Solwezi, fl. & fr. 20.ix.1930, *Milne-Redhead* 1158 (K, holotype).

Glabrous perennial herb with annual stems from a fleshy rootstock; rootstock to c. 6 × 4 cm, subglobose or ovoid; stems subfleshy, leaf-bearing or flowering, to 15 cm high, flowering separately from the leaves. Leaves sessile; stipules minute, subulate, quickly deciduous; lamina subcoriaceous, to 9 × 2 cm, oblanceolate, obtuse at the apex, cuneate at the base, entire, much smaller and often scale-like on the flowering stems. Cymes in terminal 2–3-branched umbels with primary rays to 3 cm long, each forking up to 4 times. Bracts to 2 × 1.5 cm, deltoid, apex obtuse-apiculate, base subcordate. Cyathia on peduncles to 8 mm long, 5 × 9 mm, with cup-shaped involucres; glands 5, spreading, 1.5 × 2.5–3 mm, transversely elliptic, outer margin crenulate; lobes 1.5 × 3–4 mm, transversely rectangular, margin shortly toothed and minutely ciliolate. Male flowers many: bracteoles laciniate with feathery apices; stamens 5.5 mm long. Female flower: styles 2.5 mm long, joined at the base, with widely spreading shortly bifid thickened apices. Capsule 8 × 10 mm, deeply 3-lobed; pedicel 3 mm long. Seeds 4.5 mm in diameter, globose; surface reticulately wrinkled with prominent ridges, dark brown.

Zambia. N: Mbala, fl. & fr. 11.vii.1964, *Mutimushi* 867 (K; NDO; SRGH). W: Mufulira, fl. & fr. 23.viii.1955, *Fanshawe* 2426 (K; NDO); Luanshya, 21.ii.1956, *Fanshawe* 2787 (K; NDO).
Apparently restricted to N and W Provinces of Zambia. In rocky lateritic soils in open woodland, flowering shoots appearing often after annual fires; 1220-1650 m.

49. **Euphorbia trichadenia** Pax in Bot. Jahrb. Syst. **19**: 125 (1894). —N.E. Brown in F.T.A. **6**, 1: 523 (1911); in F.C. **5**, 2: 251 (1915). Type from Angola.

Perennial herb with annual stems from a large fleshy tuberous rootstock; stems 1–several, up to 12 cm high. Leaves sessile; stipules filamentous, 0.5 mm long, quickly deciduous; lamina to 10.5 × 0.5 cm, linear-lanceolate, apiculate at the apex, cuneate at the base, entire, midrib prominent on the lower surface. Cymes in terminal 3-branched umbels, with rays to 6 cm long, not forking, or cyathia solitary. Bracts leaf-like. Cyathia on peduncles to 3 mm long, c. 5 × 10 mm, with funnel-shaped involucres; glands 5, 2 × 5 mm, transversely elliptic, inner margin raised, outer margin with 6–10 spreading finger-like processes c. 2 mm long which branch and rebranch at the tips; lobes 1.5 × 2.5 mm, transversely elliptic, margin deeply and finely toothed. Male flowers: bracteoles fan-shaped, apex laciniate, feathery; stamens

7.5 mm long, pedicels pubescent. Female flower: styles c. 3 mm long, joined to about halfway with spreading, slightly thickened, scarcely bifid apices. Capsule c. 6 × 8 mm, obtusely 3-lobed; pedicel to c. 4 mm long. Seeds 5 × 4.2 mm, ovoid with obtuse apex; surface scabrous-papillose, pale brown.

Var. **trichadenia**

Plant entirely glabrous, or rarely with a few minute hairs on the involucre, or glabrous and only minutely hispid on the stems.

Botswana. SE: Kgatleng Distr., Masama Ranch, fl. 10.xi.1978, *O.J. Hansen* 3546 (SRGH).
Also in Angola, Namibia and less commonly in northern parts of South Africa.

Var. **gibbsiae** N.E. Br. in F.T.A. **6**, 1: 524 (1911). Syntypes: Zimbabwe, Matopos (Matoppo) Hills, 1525 m, fl. x.1905, *Gibbs* 234 (BM; K, fragment); and Masvingo (Victoria) *Munro* 141 (BM; SRGH).
Euphorbia trichadenia sensu Eyles in Trans. Roy. Soc. South Africa **5**: 400 (1916), as to *Gibbs* 234 & *Munro* 141, non Pax.

Stems, cyathia and capsules pubescent.

Zimbabwe. N: Mazowe (Mazoe), U.C.R.N. Farm, fl. & fr. 2.xii.1961, *Leach* 11285 (SRGH). W: Matobo Distr., Besna Kobila Farm, 1460 m, fl. & fr. xii.1957, *Miller* 4839 (K; SRGH). C: Shurugwi (Selukwe), fl. x.1921, *Eyles* 3703 (SRGH). S: Masvingo Distr., Mutirikwi Recreational Park (Kyle National Park) H.Q., st. 17.iii.1971, *Basera* 281 (SRGH).
Not known elsewhere. In grassland; 1300–1525 m.

50. **Euphorbia grantii** Oliv. in Trans. Linn. Soc., London **29**: 144 (1875). —N.E. Brown in F.T.A. **6**, 1: 527 (1911), pro parte excl. *Hildebrandt* 2632. —S. Carter in F.T.E.A., Euphorbiaceae, part 2: 457, fig. 85/1–2 (1988). Types from Tanzania.
Euphorbia mulemae Rendle in J. Linn. Soc., Bot. **37**: 209 (1905). Syntypes from Uganda.

A few-stemmed bush or sparingly branched tree 1.5–9 m high, with a smooth but horizontally grooved grey bark; branches semi-succulent, with large prominent closely-set leaf scars. Leaves sessile, glabrous; stipules glandular, minute, deciduous; lamina to 30 × 3 cm, linear to linear-lanceolate, acuminate at the apex, rounded at the base, entire, midrib prominent on the lower surface, pale green and slightly glaucous. Cymes in terminal 3-branched umbels produced on a leafless peduncle to 10(15) cm long, with primary rays to 5(7) cm long, each forking up to 8 times, peduncle, rays and branches minutely puberulous. Bracts sessile, c. 4 × 4 cm when mature, deltoid, apex acuminate to c. 2 cm long, base subcordate, lower surface minutely puberulous at least towards the base, bracts below the umbel larger and longer. Cyathia on peduncles to 1 cm long, c. 1 × 3 cm, with barrel-shaped involucres, all parts except the upper surface of the glands minutely puberulous; glands 4 spreading, c. 4 × 8 mm, transversely elliptic, outer margin with 6–10 finger-like processes c. 8 mm long, branching several times at the tips and terminating in minute knobs, the glands yellowish-green, with reddish processes; lobes c. 3 × 5 mm, rounded, margin sharply toothed, often with a longer central tooth to 1.5 mm long, densely puberulous on both surfaces. Male flowers: bracteoles many, fan-shaped, deeply divided, feathery at the apices; stamens 14.5 mm long. Female flower: perianth obvious below the ovary, irregularly and acutely lobed with lobes to 1.5 mm long; ovary minutely puberulous; styles 1 cm long, joined to nearly halfway, with shortly bifid thickened rugulose apices. Fruit exserted on a curved pedicel to 1.5 cm long; capsule c. 13 × 17 mm, subglobose, glabrous when mature, tinged purplish. Seeds c. 6 × 5 mm, subglobose, slightly compressed laterally and obscurely 3-angled, surface minutely roughened, greyish-brown.

Zambia. N: Mbala Distr., road to Kaka Village, fl. 20.ii.1960, *Richards* 12511 (K; NDO).
Also in western Tanzania, Uganda, Burundi and Dem. Rep. Congo; recorded from the Flora Zambesiaca area only from near Mbala. In sandy soil amongst rocks in open woodland; 1675–1740 m.

Sect. 9. PSEUDEUPHORBIUM (Pax) Pax & K. Hoffm. in Engler, Pflanzenw.
Afrikas [Veg. Erde 9] **3**, 2: 163 (1921).
Sect. *Eremophyton* Subsect. *Pseudeuphorbium* Pax in Engler & Prantl, Nat.
Pflanzenfam. **3**, 5: 107 (1891).

Perennial herbs or small shrubs, with thick succulent stems simple or sparingly branched usually completely covered with large fleshy, spirally arranged tubercles bearing the leaves at their apices (see TAB. **76**, fig. B1). Leaves subsessile, entire; stipules absent. Cymes in terminal umbels forking many times, produced on slender leafy branches arising from the axils of the tubercles. Bracts sessile, ± deltoid, or leaf-like below the umbel. Cyathia with involucres bisexual; glands 4 or 5, with finger-like processes on the outer margin; lobes 5. Stamens exserted from the involucres; bracteoles included, apices feathery. Perianth of the female flower reduced to a rim below the ovary; styles with minutely bifid apices. Capsule distinctly exserted on a recurved pedicel, relatively large, subglobose. Seeds subglobose with very obtuse apex, surface obscurely wrinkled, caruncle absent.

Stems erect to 30 cm high, simple, or rhizomatous and branching underground to form
    spreading plants · 51. *monteiri*
Sparingly branched shrubs to 3 m high · 52. *wildii*

51. **Euphorbia monteiri** Hook.f. in Bot. Mag. **91**: t. 5534 (1865). —Boissier in de Candolle, Prodr. **15**, 2: 1264 (1866). —N.E. Brown in F.T.A. **6**, 1: 526 (1911). —White, Dyer & Sloane, Succ. Euphorb. **1**: 267 (1941). —Leach in Kirkia **6**: 134 (1968). Type from Angola.

Glabrous perennial, very occasionally rhizomatous, with an erect succulent tuberculate stem to 30 cm high, rarely more, to 10 cm thick, rarely branching; tubercles in 5–8 spiralled series, rhomboid to about 15 × 12 mm, scarcely prominent, bearing the leaves at their apices. Leaves subsessile; lamina to 12(21) × 1.3(3) cm, linear lanceolate, apiculate at the apex, tapering to the base, entire. Cymes in axillary and terminal 1–3-branched umbels produced on slender leafy branches to 20 cm long arising from the tubercle axils with primary rays to 8 cm long, each forking many times. Bracts sessile, c. 3 × 2–3 cm, deltoid, apex acuminate, base subcordate, the bracts below the umbel larger and more leaf-like. Cyathia on peduncles 2–5 mm long, 5 × 12 mm with cup-shaped involucres; glands 4, or 5 on the central cyathium of the umbel, c. 1.5 × 4 mm, transversely elliptic, outer margin with 3–6 finger-like processes 1–2 mm long which branch and rebranch at the tips, terminating in minute knobs, purplish-red, rarely green; lobes 2.5 × 2.5 mm, subquadrate, margin deeply toothed, minutely puberulous. Male flowers: bracteoles fan-shaped, laciniate, feathery; stamens 8 mm long. Female flower: ovary pedicellate; styles c. 4 mm long, joined to c. halfway with spreading thickened minutely bifid apices. Capsule on an erect pedicel c. 1 cm long, 8 × 11 mm, subglobose, 3-lobed, with 6 longitudinal ridges when dry. Seeds 4.2 × 3.8 mm, subglobose, apex very obtuse, surface obscurely wrinkled, yellowish-grey.

Subsp. **monteiri** —O.B Miller in J S. African Bot **18**: 44 (1952). —Leach in Kirkia **6**: 135, photos 1–4 (1968).
    Euphorbia marlothii Pax in Bot. Jahrb. Syst. **10**: 36 (1888). Type from Namibia.
    Euphorbia longibracteata Pax in Bull. Herb. Boissier **6**: 741 (1898). Type from Namibia.
    Euphorbia baumii Pax in Bull. Herb. Boissier, sér. 2, **8**: 636 (1908). Type from Angola.

Stems simple or branching usually only when damaged; tubercles closely spiralled; bracts c. 3 × 2 cm, deltoid.

Botswana. N: Kgwebe Hills, fl. 5.i.1898, *Mrs Lugard* 87 (K); fr. i.1897, *Maj. Lugard* 247 (K). SW: Kobe Pan (Koobies, or Kobis), fl. 28.x.1861, *Baines* s.n. (K). **Zimbabwe**. W: Hwange Distr., 3 km south of Beacon 389 on Botswana border, cult. Harare, fl. 1.xii.1970, *Biegel* 3459 (SRGH).
    Mainly distributed in Namibia, occurs also in southern Angola. On sandy soils in wooded grassland; c. 1005–1070 m.

Subsp. **ramosa** L.C. Leach in Kirkia **6**: 138, photos 5–7 (1968). Type from South Africa (Transvaal).

Euphorbia monteiri sensu Phillips in S. African J. Sci. **16**: 430 (1920). — sensu Pole Evans (ed.), Fl. Pl. South Africa **6**: pl. 218 (1926), non Hook.f. sensu stricto.

As for subsp. *monteiri*, but occasionally rhizomatous, branching below ground to form spreading plants; tubercles loosely spiralled; bracts c. 3 × 3 cm broadly deltoid.

Botswana. SE: Serowe, *Mogg* 24561 (PRE).
Native to the Transvaal, and according to Leach, *Mogg* 24561 represents a discrete population in Botswana (I have not seen the specimen). Sandy soil amongst rocks; c. 1000 m.

52. **Euphorbia wildii** L.C. Leach in Kirkia **6**: 139, photos 8–11 (1968); in Kirkia **10**: 293 (1975)*.
TAB. **76**, fig. B. Type: Zimbabwe, Mazowe Distr., Mvurwi Range (Umvukwes), Ruorka Ranch, 16.xii.1952, *Wild* 3981 (SRGH, lectotype).
Euphorbia monteiri sensu Wild in Kirkia **5**: 77 (1965) non Hook.f.

Stout succulent shrub to 3 m high, glabrous, sparingly branched; branches to 5 cm in diameter, tuberculate; tubercles tightly spiralled in 5–8 or more series, rhomboid to about 15 × 15 mm, very prominent, bearing the leaves at their apices. Leaves subsessile; lamina to 12 × 4 cm, obovate-lanceolate, acute and apiculate at the apex, tapering to the base, entire. Cymes in axillary and terminal 1–5-branched umbels produced on slender leafy branches to 10 cm long, arising from the tubercle axils, with primary rays to 7 cm long, each forking many times and the dry remains persisting towards the apex of the main branches. Bracts sessile, c. 3.5 × 4 cm, deltoid, apex acuminate, base subcordate, the bracts below the umbel larger and more leaf-like. Cyathia on peduncles c. 2 mm long, 7 × 15 mm with cup-shaped involucres; glands 4(5), c. 2 × 3.5 mm, transversely elliptic, outer margin with 2–7 finger-like processes 2–4 mm long with swollen, sometimes scarcely bifid tips, green; lobes 3 × 3 mm, subquadrate, margin very deeply toothed, puberulous. Male flowers: bracteoles fan-shaped, laciniate, feathery; stamens 8 mm long. Female flower: ovary pedicellate; styles c. 4 mm long, joined to halfway, with spreading slightly thickened apices. Fruit exserted on a curved pedicel c. 1 cm long; capsule to 10 × 13 mm, subglobose, 3-lobed. Seeds 4 × 3.5 mm, ovoid, apex very obtuse, surface very obscurely wrinkled, brown.

Zimbabwe. N: Zvimba Distr., west side of Mvurwi (Umvukwes) Range, c. 3 km north of Kildonan, fl. 24.x.1959, *Leach* 9498 (K, SRGH); Mutorashanga (Mutoroshanga), fl. 17.ii.1966, *Leach, Simon & Scott* 13196 (BM; COI; K; SRGH).
Endemic to the Mvurwi (Umvukwes) Mts. On rocky slopes of serpentine hills with a high chrome content; 1480–1700 m.

Sect. 10. MEDUSEA (Haw.) Pax & K. Hoffm. in Engler, Pflanzenw. Afrikas [Veg. Erde 9] **3**, 2: 161 (1921).
Medusea Haw., Pl. Succ.: 131 (1812).

Succulent dwarf perennials with a large fleshy tapering root merging into a thick, much abbreviated sometimes underground main stem, rarely with several stems clustered, usually producing many densely packed spirally arranged succulent branches radiating from the apical growing point situated ± at ground level; stems and branches completely covered with spirally arranged fleshy tubercles bearing leaves at their apices. Leaves small, usually linear, deciduous; stipules apparently absent. Cyathia solitary, on short often leafy peduncles arising in the axils of the tubercles; peduncles occasionally persistent and becoming woody; bracts sessile, shorter than the involucres. Involucres bisexual (plants sometimes dioecious in South Africa); glands 5, entire, crenulate, lobed, or decorative with finger-like processes on the outer margin; lobes 5. Stamens clearly exserted from the involucre;

---

* *Euphorbia wildii* L.C. Leach was validly published in Kirkia **6**: 139 (1968) despite the mention there of duplicate specimens in different herbaria, since a single gathering was cited, see Art. 37.2 of Saint Louis Code. Leach is interpreted as having later lectotypified the name in Kirkia **10**: 293 (1975) when he republished it giving the SRGH specimen as the holotype, see Art. 9.8 of Saint Louis Code.

bracteoles included, or occasionally with feathery tips exserted. Perianth of the female flower reduced to a rim below the ovary. Capsule obtusely 3-lobed, sessile. Seeds ovoid, often rugulose, without a caruncle.

Branches c. 4 mm in diameter, in clusters at ground level, of c. 6–9 from the apices of stems to
    c. 2 cm in diameter; leaves 1.5–3 cm long · 53. *duseimata*
Branches c. 10 mm in diameter, numerous, radiating from the central growing point of the
    stem up to c. 8 cm in diameter; leaves up to 1 cm long · · · · · · · · · · · · · · · 54. *maleolens*

53. **Euphorbia duseimata** R.A. Dyer in Fl. Pl. South Africa **14**: t. 530 (1934). —White, Dyer & Sloane, Succ. Euphorb. **1**: 410 (1941). Syntypes: Botswana, c. 160 km northwest of Molepolole, *G.J. de Wijn* in *NH* 12426 (PRE, syntype); Seletse (Seletsi), 80 km south of Molepolole, *Knobel* s.n. in *NH* 15916 (PRE, syntype).

Dwarf perennial; rootstock tuberous often greatly elongated, 1.25–2.5 cm in diameter; stems 1–3, occasionally more, produced underground, to 7–10 cm long and 0.5–2 cm in diameter; branches 6–9, in clusters from each stem apex, produced at about ground level, 4–6 cm long, 2.5–4 mm in diameter, cylindric and covered in small tubercules, rebranching on older plants and eventually forming rounded clumps to c. 50 cm in diameter; tubercules elongated, 5–8 mm long with a prominent white apex c. 1 mm wide. Leaves 5–30 × 2 mm, linear, fleshy, crowded towards the branch tips. Cyathia produced towards the branch tips on non-persistent peduncles 5–10 mm long; bracts 3–8, obovate-oblong, scale-like, deciduous. Cyathia 3 × 4 mm with cup-shaped involucres, glabrous; glands 2–2.5 × 1 mm, transversely elliptic, outer margins with 2–5 entire or bifid processes to 1 mm long, greenish-white often spotted with red; lobes 1 × 1.5 mm, rounded, ciliate. Male flowers: bracteoles 2 mm long, laciniate; stamens 3 mm long. Female flower: ovary glabrous; styles united for 2–2.5 mm, stigmas short, spreading. Capsule c. 5 mm in diameter, obtusely lobed, sessile. Seeds 3.5 × 2.8 mm, ovoid with truncate base and pointed apex, with scarcely prominent warts, greyish-brown.

Botswana. SW: 19 km west of Manatse, 99 km east of Axade, fr. 24.iv.1984, *Woollard* 1386 (SRGH). SE: 20 km northwest of Molepolole, fl. & fr. 1.xii.1954, *Codd* 8924 (K; PRE).
Also in South Africa (North-West Prov.). Sandy soils in open bushland; c. 1035–1100 m.

54. **Euphorbia maleolens** Phillips in Fl. Pl. South Africa **12**: t. 459 (1932). —White, Dyer & Sloane, Succ. Euphorb. **1**: 407 (1941). Type from South Africa (Northern Prov.).

Dwarf perennial with an evil smelling latex; rootstock tuberous, thick, tapering, merging into the stem below ground to form a body to c. 20 cm long, 3–8 cm in diameter, with c. 2.5 cm of the stem above ground, rarely divided into 2 or more; stem apex truncate, covered with prominent rhomboid tubercles 10–15 × 5–12 mm; branches radiating and ascending from around the apical growing point, to 8–20 cm long, 1 cm in diameter with small prominent, white-tipped tubercles. Leaves clustered towards branch tips, to 10 × 2 mm, linear-lanceolate, entire, longitudinally folded, sessile. Cyathia produced towards branch tips, with peduncles to 8–12 mm long, persistent and becoming woody; bracts 4–6, scale-like, deciduous. Cyathia c. 2.5 × 3.5 mm with cup-shaped involucres; glands widely separated, c. 1.7 × 2 mm, subquadrate, dark green, outer margins with 2–4 slender, subulate, yellow processes 1–2 mm long; lobes c. 1 × 1.5 mm, transversely oblong, toothed, ciliate. Male flowers: bracteoles 3 mm long, laciniate; stamens 4.5 mm long. Female flower: styles united for 1.5 mm, stigmas stout, recurved. Capsule c. 5 × 6 mm, obtusely lobed, green with red-brown lines along the sutures, sessile. Seeds c. 3.2 × 2.5 mm, ovoid with truncate base and pointed apex, smooth.

Botswana. SE: 5 km west of Ramotswa Station on Thamaga road, fr. 21.viii.1977, *O.J. Hansen* 3163 (K; SRGH).
Also in South Africa (Northern Prov.). Sandy soils in mixed bushveld; c. 1050–1130 m.
Said to occur in Zimbabwe (west of Bulawayo), but I have seen no specimen or other evidence in support of this.

Sect. 11. LYCIOPSIS Boiss. in de Candolle, Prodr. **15**, 2: 97 (1862). —S. Carter in F.T.E.A., Euphorbiaceae, part 2: 461 (1988).

Shrubs or small trees, or woody herbs, roots sometimes fleshy and tuberous. Leaves shortly petiolate; stipules modified as conspicuous glands, often sharply pointed. Cymes terminal or axillary, in 2–7-branched umbels with rays forking up to 3 times, or the cyathia solitary; bracts scarious or leafy. Cyathia with involucres bisexual and 5(8) entire, saucer-shaped to funnel-shaped or 2-lipped involucral glands; lobes 5. Stamens and feathery tips of the bracteoles exserted from the involucre. Perianth of the female flower reduced to a rim below the ovary; styles thickened, spreading and grooved on the upper (inner) surface. Capsule subsessile, shallowly 3-lobed and pubescent. Seeds subglobose (in the Flora Zambesiaca area), smooth, without a caruncle.

1. Perennial herb with a large tuberous rootstock; stems erect to c. 10 cm tall, or decumbent to 30 cm long · 55. *oatesii*
 – Woody shrubs or small trees, to 3–4 m high · 2
2. Branching alternate, with spine-tipped branchlets of uniform length at right angles to the branches; leaves glabrous · 56. *cuneata*
 – Branching trichotomous, at least at the branch apices; leaves pubescent on the lower surface, at least when young · 57. *matabelensis*

55. **Euphorbia oatesii** Rolfe in Oates, Matabeleland Victoria Falls, ed. **2**, appendix V: 408 (1889). —N.E. Br. in F.T.A. **6**, 1: 522 (1911). —Eyles in Trans. Roy. Soc. South Africa **5**: 400 (1916). TAB. **77**, fig. A. Syntypes: Zambia, Chilanga, fl. 9.ix.1909, *Rogers* 8466 (K). Zimbabwe, Matabeleland, fl. 1878, *Oates* (K); 160 km northeast of Bulawayo, *Rand* 218 (not traced).

Perennial herb; rootstock large tuberous, producing several woody subterranean stems up to 1.5 cm thick branching at ground-level; branches erect, up to c. 10(20) cm high, or decumbent to 30 cm long, usually woody or sometimes herbaceous when produced after annual fires, shortly pubescent with crisped hairs. Leaves with a puberulous petiole to 5 mm long; stipules small, glandular, reddish, lamina to 70 × 9 mm, usually much less, linear-lanceolate, acute apiculate at the apex, rounded at the base, entire, midrib prominent beneath, margin and midrib beneath usually pubescent especially towards the base. Cymes reduced to solitary cyathia terminating leafy branches. Cyathia on pubescent peduncles 1–3(5) cm long, 4.6 × 7 mm, with cup-shaped involucres, pubescent with spreading crisped hairs; glands 5(6), distant, pink, c. 1.5 × 2.2 mm, transversely elliptic, markedly 2-lipped to funnel-shaped with a transverse opening, margins entire; lobes c. 1.75 × 2.5 mm, transversely rectangular, margin deeply toothed, ciliate. Male flowers many: bracteoles laciniate, feathery; stamens 4.5 mm long, with pedicels usually pubescent at the apex. Female flower: styles 2 mm long, pubescent, joined to halfway, with spreading recurved shortly bifid apices, channelled on the upper (inner) surface. Capsule subsessile, shallowly 3-lobed, 5 × 7 mm, densely pubescent. Seeds 3.5 × 3.2 mm, subglobose, very obtusely pointed at the apex, smooth, brown speckled.

Zambia. B: Kaoma Distr., 80 km west of Kafue Hoek pontoon on road to Kaoma (Mankoya), fl. & fr. 7.xi.1959, *Drummond & Cookson* 6219 (K; SRGH). W: Chingola, fl. & fr. 8.x.1954, *Fanshawe* 1591 (K; SRGH). C: Lusaka Distr., Chakwenga Headwaters, 160–205 km east of Lusaka, fl. 29.ix.1963, *E.A. Robinson* 5673 (K; SRGH). S: Namwala Distr., 7 km from Namwala on Ngoma road, fl. & fr. 8.xii.1962, *van Rensburg* 1041 (K; SRGH). **Zimbabwe**. N: Hurungwe Distr., Musukwi (Msukwe) R., Kanyanga Stream, fl. & fr. 18.xi.1953, *Wild* 4175 (K; SRGH). W: Binga Distr., Chebira Hot Spring, fl. x.1948, *Whellan* 390 (K; MAL; SRGH). C: Kadoma Distr., Silverstream Ranch, south of Ngesi, fl. 16.i.1962, *Wild* 5609 (K; SRGH).
Not known elsewhere. In sandy soil in grassland and open *Brachystegia* woodland; 850–1370 m.

56. **Euphorbia cuneata** Vahl, Symb. Bot., part 2: 53 (1791). —N.E. Brown in F.T.A. **6**, 1: 545 (1911). —S. Carter in Kew Bull. **35**: 423 (1980); in F.T.E.A., Euphorbiaceae, part 2: 466 (1988). Type from N Yemen.

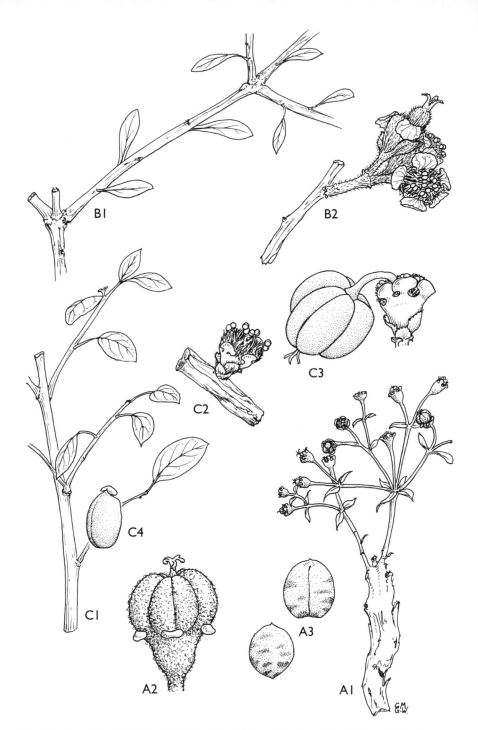

Tab. 77. A. —EUPHORBIA OATESII. A1, habit (× ²⁄₃), from *E.A. Robinson* 896; A2, sessile capsule (× 3), from *Wild* 1286; A3, seeds (× 4), from *Greenway & Brenan* 8092. B. — EUPHORBIA MATABELENSIS. B1, leafy branch (× ²⁄₃), from *Brummitt* 10248; B2, cyme, with developing fruit (× 3), from *Chase* 1726. C. —EUPHORBIA ESPINOSA. C1, leafy branch (× ²⁄₃), from *Fanshawe* 6315; C2, cyathium (× 3), from *Angus* 151; C3, exserted capsule (× 3); C4, seed (× 3), C3 & C4 from *Fanshawe* 6661. Drawn by Christine Grey-Wilson.

Subsp. **cuneata**
> *Euphorbia cuneata* var. *carpasus* Boiss. in de Candolle, Prodr. **15**, 2: 97 (1862). Types from Eritrea and Zanzibar.

Woody shrub or small tree to 4 m high with brownish-grey peeling bark; branches with alternating spine-tipped branchlets to 10 cm long, spreading at right angles, pubescent when young. Leaves alternate on young growth, fasciculate on older branches, with a petiole to 5 mm long; stipules glandular, minute and sharply pointed; lamina to 4.5 × 1.5 cm, cuneate-spathulate, rounded to emarginate at the apex, cuneate at the base, entire, glabrous. Cymes clustered in terminal 2–5-branched umbels with rays to 7 mm long, not forking. Bracts minute, 1–3 mm long, leaf-like, puberulous. Cyathia 4–5 × 7 mm with broadly funnel- to cup-shaped involucres, minutely puberulous; glands 5, to c. 2.5 mm in diameter, saucer-shaped, yellow; lobes c. 1.5 × 2.5 mm, rounded with deeply fringed margin. Male flowers: bracteoles ligulate with feathery apices; stamens c. 4.5 mm long. Female flower: ovary subsessile, puberulous; styles 2 mm long, joined to ± halfway with recurved thickened bifid apices, grooved on the upper surface. Capsule to 7 × 8 mm, obtusely 3-lobed, pubescent. Seeds c. 2.5 mm in diameter, subglobose, smooth, reddish-brown, faintly speckled.

Mozambique. M: Marracuene, fl. 2.x.1957, *Barbosa & Lemos* 7892 (COI; K; LISC).
From the Arabian Peninsula and Jebel Elba southwards through the Red Sea Hills of Sudan, eastern Ethiopia, and coastal regions of Somalia, Kenya and Tanzania; in Mozambique it is known only from Maputo District where it has escaped from cultivation as a hedge plant; c. 80–90 m.
Several subspecies are known in East Africa, and other undescribed taxa occur in Somalia and Ethiopia, differing from the typical subspecies in bark colour, pubescence, branching of the cymes or shape of the involucral glands.

57. **Euphorbia matabelensis** Pax in Ann. Naturhist. Mus. Wien **15**: 51 (1900). —N.E. Brown in F.T.A. **6**, 1: 546 (1911). —S. Moore in J. Linn. Soc., Bot. **40**: 190 (1911). —Eyles in Trans. Roy. Soc. South Africa **5**: 398 (1916). —White, Dyer & Sloane, Succ. Euphorb. **1**: 89 (1941). —Suessenguth & Merxmüller, [Contrib. Fl. Marandellas Distr.] Proc. & Trans. Rhod. Sci. Ass. **43**: 84 (1951). —Wild in Clark, Victoria Falls Handb.: 148 (1952). —Brenan in Mem. New York Bot. Gard. **9**: 67 (1954). —F. White, F.F.N.R.: 198 (1962). —S. Carter in Kew Bull. **40**: 823 (1985); in F.T.E.A., Euphorbiaceae, part 2: 469 (1988). TAB. **77**, fig. B. Type. Zimbabwe, Matabeleland, *Penther* 944 (BM, isotype).
> *Euphorbia jaegeriana* Pax in Bot. Jahrb. Syst. **43**: 87 (1909). Type from Tanzania.
> *Euphorbia currori* N.E. Br. in F.T.A. **6**, 1: 545 (1911). Type from Angola.
> *Euphorbia inelegans* N.E. Br. in F.T.A. **6**, 1: 547 (1911). Type from Tanzania.

Woody shrub or usually a small tree to 3(5) m high, deciduous, with peeling greyish-brown bark; young branches densely puberulous, glabrescent; at least some of the branching pattern trichotomous, especially at the branch apices, branchlets often spine-tipped. Leaves alternate or more usually fasciculate, with petioles 1–5 mm long; stipules glandular, minute, linear; lamina to 5.5 × 2.3 mm, oblanceolate to obovate, obtuse to rounded at the apex, cuneate at the base, entire, lower surface pubescent when young, upper surface scarcely so, glabrescent. Cymes terminal, usually on short axillary branches, in 3–7-branched umbels with rays to 1 cm long, 0–2-forked. Bracts leaf-like, c. 5 mm long, often yellowish-green. Cyathia subsessile, or the central one of the umbel on a peduncle to 5 mm long, densely pubescent, c. 3.5 × 6 mm, with cup-shaped involucres; glands 5, spreading, 2–2.5 mm wide, rounded, shallowly saucer-shaped, yellow; lobes c. 1.5 × 2 mm, rounded, deeply fringed. Male flowers: bracteoles fan-shaped, deeply divided with feathery apices; stamens c. 4.5 mm long. Female flower: ovary subsessile, densely pubescent; styles 2 mm long, joined at the base, with widely spreading flattened rugose bifid apices. Capsule slightly exserted on an erect pedicel to 5 mm long, c. 7 × 8 mm, obtusely 3-lobed, densely pubescent. Seeds 3.5 mm in diameter, globose, smooth, brown and obscurely speckled.

Caprivi Strip. 41 km west of Katima Mulilo, fl. 17.ii.1969, *de Winter* 9205 (K; PRE). **Botswana**. N: Chobe, fl. 28.vii.1950, *Robertson & Elffers* 62 (K; PRE; SRGH). **Zambia**. N: Samfya, fl. 3.ix.1953, *Fanshawe* 266 (BR; K; NDO). W: 2 km from Ndola, st. 30.xii.1961, *Symoens* 9127 (K). C: Kafue, fl. & fr. 2.iii.1963, *van Rensburg* 1561 (K; SRGH). E: Chipata Distr., Luangwa Valley, near Jumbe, fl. 16.v.1968, *Astle* 5145 (K; NDO; SRGH). S: Choma Distr., Choma National Forest

(Siamambo Forest Reserve), near Choma, fl. 29.vii.1952, *Angus* 85 (K). **Zimbabwe**. N: Hurungwe Distr., Urungwe National Park, 300.5 km from Harare on Chirundu road, fl. & fr. 20.ii.1981, *Philcox, Leppard & Dini* 8788 (K; SRGH). W: Hwange Distr., Victoria Falls, east of Chinotimba, 880 m, fl. & fr. 26.ix.1978, *Ncube* 43 (K; SRGH). C: Harare Distr., Domboshawa, fl. 16.ii.1947, *Wild* 1655 (K; SRGH). E: Mutasa Distr., Honde Valley, fl. & fr. 24.viii.1949, *Chase* 1726 (K; SRGH). S: Gutu Distr., Devure Purchase Area, fl. x.1959, *R.M. Davies* D2602 (SRGH). **Malawi**. N: Rumphi Distr., north of Rumphi Boma, fr. 17.viii.1972, *Brummitt & Patel* 12904 (K; SRGH). C: Salima Distr., Senga Bay, fl. & fr. 25.iv.1971, *Pawek* 4714 (K; MAL; SRGH). S: Machinga Distr., Munde Hill, saddle, fl. 30.iv.1982, *Patel* 885 (BR; K; MAL). **Mozambique**. N: Sanga Distr., Matunde to Messinge R., fl. 13.ix.1934, *Torre* 581 (COI; LISC). T: Cahora Bassa, planalto do Songo, junto ao campo de aviação, fl. 12.vi.1971, *Torre & Correia* 18734 (K; LISC). MS: Rotanda, at km 4 on Mavita road, fl. 30.x.1965, *Torre & Pereira* 12615 (LISC).

Also in Tanzania and southern Kenya, and just extends into southern Angola. In open deciduous woodlands, in sandy soils and often on rocky outcrops and hillsides; 450–1525 m.

In drier situations, towards the Tanzanian border, the leaves are often more pubescent, and the branching is sturdier and more obviously trichotomous.

Sect. 12. ESPINOSAE Pax & K. Hoffm. in Engler, Pflanzenw. Afrikas [Veg. Erde 9]
   3, 2: 149 (1921). —S. Carter in F.T.E.A., Euphorbiaceae, part 2: 470 (1988).

Woody shrubs, sometimes scandent. Leaves shortly petiolate; stipules modified as conspicuous glands. Cyathia solitary, axillary, subsessile, surrounded at the base by a cluster of small leaf-like or scarious bracts. Involucres bisexual, with 5 entire, spreading glands; lobes 5. Stamens shortly exserted from the involucre; bracteoles included. Ovary subtended by an obvious 3-lobed perianth; styles joined at the base, with spreading bifid apices. Capsule well-exserted on a reflexed pedicel, relatively large, deeply 3-lobed, glabrous. Seeds ovoid, slightly dorsi-ventrally compressed, smooth, with a cap-like caruncle.

Leaves elliptic, to 4.5(6) × 3 cm; petiole to 12 mm long · · · · · · · · · · · · · · · · · · · 58. *espinosa*
Leaves lanceolate, to 1.5(3.5) × 0.5(1.3) cm; petiole 1–3 mm long · · · · · · · · · 59. *guerichiana*

58. **Euphorbia espinosa** Pax in Bot. Jahrb. Syst. **19**: 120 (1894). —N.E. Brown in F.T.A. **6**, 1: 547 (1911). —Eyles in Trans. Roy. Soc. South Africa **5**: 398 (1916). —White, Dyer & Sloane, Succ. Euphorb. **1**: 91 (1941). —F. White, F.F.N.R.: 198 (1962). —S. Carter in F.T.E.A., Euphorbiaceae, part 2: 470 (1988). TAB. **77**, fig. C. Type from Tanzania.
Euphorbia gynophora Pax in Bot. Jahrb. Syst. **34**: 374 (1904). Type from Tanzania.
Euphorbia nodosa N.E. Br. in F.T.A. **6**, 1: 548 (1911). Types from Angola.

Erect or occasionally scandent woody shrub to 3 m high, with a smooth shiny bark; young branchlets pubescent. Leaves with a petiole to 12 mm long, usually pubescent on young leaves, with long spreading hairs; stipules glandular, conspicuous, reddish; lamina to 4.5(6) × 3 cm, elliptic, rounded and minutely apiculate at the apex, abruptly cuneate at the base, entire, midrib prominent beneath, young leaves pubescent towards the base. Cymes reduced to solitary, axillary subsessile cyathia, surrounded at the base by 4–8 scarious bracts, c. 2 × 1.5 mm, with finely ciliate margins. Cyathia c. 3 × 5 mm, with funnel-shaped involucres; glands 5, c. 1.5 × 2.5 mm, transversely elliptic, yellowish-green, touching; lobes c. 1 mm long, subquadrate, margin denticulate. Male flowers: bracteoles deeply laciniate, margins ciliate; stamens 3.5 mm long. Female flower: ovary glabrous, subtended by an obvious 3-lobed perianth; styles 2 mm long, joined for one-third, with spreading bifid apices. Capsule to 8.5 × 10 mm, deeply 3-lobed with a slightly depressed apex, green often tinged with purple, exserted on a reflexed, usually puberulous, pedicel to 10 mm long. Seeds c. 5.5 × 4.5 mm, ovoid, slightly dorsi-ventrally compressed, smooth, pale grey; caruncle cap-like, 2.5 mm wide, yellow.

Caprivi Strip. Opposite Andara Mission Station, st. 21.ii.1956, *de Winter & Marais* 4785 (K; PRE). **Zambia**. B: Sesheke Distr., Machili, fl. 22.ii.1961, *Fanshawe* 6315 (K; NDO; SRGH). C: Lusaka Distr., Iolanda Waterworks, Muchuto R. Gorge, st. 4.xii.1996, *M.G. & P.E. Bingham* 11237 (K). E: Lundazi, fr. 18.viii.1965, *Fanshawe* 9287 (K; NDO). S: Choma Distr., near Ngonga (Ngongo) R. where it crosses Choma–Namwala road, fl. 3.viii.1952, *Angus* 151 (K). **Zimbabwe**. N: Gokwe Distr., Sengwa Research Station, fl. 5.x.1968, *N.H.G. Jacobsen* 240 (K; SRGH). W: Bulilima Mangwe Distr., Plumtree, fl. xii.1954, *Meara* 84B (K; SRGH). C: Chegutu Distr., Poole

Farm, fr. 31.vii.1955, *R.M. Hornby* 3375 (BR; K; SRGH). E: Mutare Distr., Odzi River, Dice-box Farm, fl. & fr. 22.viii.1949, *Chase* 1722 (COI; K; SRGH). S: Buhera Distr., west bank of Save River above Birchenough Bridge, fl. 18.viii.1961, *Chase* 7529 (K; SRGH). **Malawi**. N: 16 km northwest of Rumphi, fl. 24.iv.1977, *Pawek* 12619 (BR; K; MAL; SRGH). S: Shire Highlands (Manganja Hills), Magomero, fl. 16.ix.1861, *Meller* s.n. (K).

Also in Tanzania and southern Kenya. In mopane, miombo and other deciduous woodlands; 300–1400 m.

59. **Euphorbia guerichiana** Pax in Bot. Jahrb. Syst. **19**: 143 (1894). —N.E. Brown in F.C. **5**, 2: 270 (1915). —White, Dyer & Sloane, Succ. Euphorb. **1**: 92 (1941). —Merxmüller, Prodr. Fl. SW. Afrika, fam. part 67: 26 (1967). Type from Namibia.

Euphorbia commiphoroides Dinter, Deut. Südw. Afrika: 90 (1909). —N.E. Brown in F.T.A. **6**, 1: 543 (1911). Types from Namibia.

Euphorbia frutescens N.E. Br. in F.C. **5**, 2: 270 (1915). —White Dyer & Sloane, Succ. Euphorb. **1**: 94 (1941). Type from Namibia.

Woody shrub or small tree to 3 m high with a papery bark; branches reddish-brown. Leaves with a petiole 1–3 mm long; stipules glandular, triangular, minute, dark brown; lamina 1.5–3.5 × 0.5–1.3 cm, obovate to lanceolate, rounded at the apex, cuneate at the base, entire, minutely pubescent especially beneath, rarely almost glabrous. Cymes reduced to solitary axillary subsessile cyathia on dwarf leafy branchlets; bracts scarious, 1 × 1 mm, rounded, margins ciliate. Cyathia 2.5 × 3.5 mm with cup-shaped involucres, minutely pubescent, at least at the base; glands 5, 1 × 1.5 mm, transversely elliptic, spreading; lobes 0.5 mm long, rounded, margin denticulate, ciliate. Male flowers: bracteoles deeply laciniate, apices ciliate; stamens 3.5 mm long. Female flower: ovary glabrous, subtended by an obvious 3-lobed perianth; styles 2.5 mm long, joined to halfway with spreading bifid apices. Capsule c. 9 × 9 mm, deeply 3-lobed, minutely pubescent, exserted on a reflexed pedicel 6–9 mm long. Seeds 4.5 × 3.5 mm, ovoid with obtusely pointed apex, slightly dorsi-ventrally compressed, smooth, grey, speckled; caruncle 2.3 mm in diameter, cap-like.

Botswana. SE: 8 km east of Letlhakane (Lothlekane), fl. 24.iii.1965, *Wild & Drummond* 7259 (K; SRGH). **Zimbabwe**. S: Gwanda Distr., Native Area 'G', Marimore road, fl. 18.xii.1956, *Davies* 2354 (K; SRGH); Beitbridge, fr. 7.i.1961, *Wild* 5307 (K; SRGH).

Occurs extensively in South Africa from the Northern Province westwards to the Northern Cape, and into Namibia and southern Angola. In dry deciduous woodland; c. 560 m.

Sect. 13. TIRUCALLI Boiss. in de Candolle, Prodr. **15**, 2: 94 (1862).

Trees or shrubs with copious latex; branches fleshy, green, cylindrical, rarely woody, alternate or often in whorls. Leaves sessile, entire, small, lanceolate and quickly deciduous; stipules modified as glands or absent but with the leaf scar becoming conspicuously calloused and apparently glandular. Cyathia in terminal umbellate cymes, with rays simple or branching dichotomously; bracts scarious or leaf-like, sessile, usually quickly deciduous. Involucres bisexual or sometimes unisexual; glands 4(5) or 5–8 on the larger central primary cyathium, entire; lobes 5. Stamens and feathery apices of the bracteoles clearly exserted from the involucre. Perianth of the female flower reduced to a rim below the ovary or distinctly 3-lobed; styles joined at the base, with spreading bifid apices. Capsule exserted on a reflexed pedicel, subglobose. Seeds ovoid, with a caruncle.

Branches with longitudinal striations; cyathia in dense clusters · · · · · · · · · · · · · · 60. *tirucalli*
Branches smooth; cyathia in umbels with 4–8 simple rays · · · · · · · · · · · · · · · · 61. *gossypina*

60. **Euphorbia tirucalli** L., Sp. Pl.: 452 (1753). —Boissier in de Candolle, Prodr. **15**, 2: 96 (1862). —Hiern, Cat. Afr. Pl. Welw. **1**: 949 (1900). —N.E. Brown in Bull. Misc. Inform., Kew **1914**: 94 (1914); in F.C. **5**, 2: 293 (1915). —White, Dyer & Sloane, Succ. Euphorb. **1**: 101 (1941). —F. White, F.F.N.R.: 199 (1962). —Leach in Kirkia **9**: 69, photos (1973). —S. Carter in F.T.E.A., Euphorbiaceae, part 2: 471 (1988). Lectotype from Sri Lanka.

Euphorbia rhipsaloides N.E. Br. in F.T.A. **6**, 1: 556 (1911). Type from Angola.

Euphorbia media N.E. Br. in F.T.A. **6**, 1: 556 (1911). Syntypes: Malawi, Karonga, fl. i.1888, *Scott* s.n. (K, syntype); and numerous other syntypes from East Africa.

Euphorbia media var. *bagshawei* N.E. Br. in F.T.A. **6**, 1: 556 (1911). Types: Mozambique, Tete, fl. xi.1859, *Kirk* s.n. (K, syntype); and other syntypes from East Africa.
Euphorbia scoparia N.E. Br. in F.T.A. **6**, 1: 557 (1911). Types from Sudan and Ethiopia.

Spineless succulent densely branched often apparently dioecious shrubs to 4 m or trees to 7 m high, with a copious irritant white to yellowish latex. Branchlets brittle terete succulent, c. 7 mm thick, often produced in whorls, green with longitudinal fine striations and very small leaf scars, the extreme tips of young leafy branchlets sparsely tomentose with curled brown hairs. Leaves few, present only at the tips of young branchlets and quickly deciduous, subsessile; stipules glandular, minute, dark brown; lamina fleshy to 15 × 2 mm, linear-lanceolate. Cymes congested, 2–6 at the branchlet apices, each forking 2–4 times, with rays less than 1 mm long, producing a dense cluster of cyathia developing only male flowers, or occasionally a few female flowers also present, or cyathia fewer and only female flowers developing, the whole cyme ± glabrous, or tomentose with curled brown hairs especially on the involucres and lobes; bracts c. 2 × 1.5 mm, rounded, ± sharply keeled, usually glabrous except on the margin. Cyathia subsessile, c. 3 × 4 mm, with cup-shaped involucres; glands 5, 0.5 mm in diameter subcircular to 1.5 × 2 mm transversely elliptic, bright yellow; lobes c. 0.5 mm long, triangular. Male involucres: bracteoles linear with plumose apices; stamens 4.5 mm long; an aborted female flower is occasionally present. Female involucres: bracteoles present and occasionally also a few male flowers; female perianth distinctly 3-lobed below the tomentose ovary, with lobes 0.5 mm long; styles 2 mm long, joined at the base, with thickened deeply bifid recurved apices. Capsule glabrescent, c. 8 × 8.5 mm, subglobose, exserted on a tomentose pedicel to 10 mm long. Seeds 3.5 × 2.8 mm, ovoid, smooth, buff speckled with brown and with a dark brown ventral line; caruncle 1 mm across.

Botswana. N: Maun, fl. (female), 27.xi.1977, *P.A. Smith* 2111 (K; SRGH). SE: Serowe, st. iv.1967, *Mitchison* A35 (K). **Zambia**. E: Changwe, between Petauke and Mwape, fl. (male), 16.xii.1958, *Robson* 961 (K; SRGH). S: Mazabuka Distr., Kafue Flats, fl. (male), 7.x.1930, *Milne-Redhead* 1257 (K); Mumbwa Distr., Shakatende, st. 1.viii.1963, *van Rensburg* 2384 (K; SRGH). **Zimbabwe**. N: Mount Darwin Distr., 3 km from Mukumbura R., towards escarpment, st. 23.i.1960, *Phipps* 2415 (K; SRGH). W: Tsholotsho Distr., 40 km south of Tsholotsho (Tjolotjo), st. vii.1971, *Leach & Cannell* 14792 (K; SRGH). C: Harare, Highlands, st. 10.iv.1970, *Leach* 14463 (K; LISC; SRGH). E: Mutare Distr., foothills of Vumba Mt., fl. (male), 1.xii.1953, *Chase* 5161 (K; SRGH). S: Mberengwa Distr., Ngezi River by Mt. Buhwa, fr. 1.xi.1973, *Biegel, Pope & Gosden* 4349 (K; SRGH). **Malawi**. N: Mzimba Distr., Euthini (Eutini), fl. (female), 27.xi.1976, *Pawek* 11980 (K; MAL; SRGH). S: Zomba Distr., Nakaronje Hill, fl. (male), 8.x.1978, *Hargreaves* 507 (MAL). **Mozambique**. Z: Milange Distr., Ruo River at Zoa Falls, st. 16.viii.1971, *Leach & Royle* 14815 (K; LISC; SRGH). T: Cahora Bassa Distr., Chicoa, fl. (male), 25.ix.1942, *Mendonça* 413 (SRGH). GI: Govuro Distr., Save R., 16 km south of Nova Mambone (Mambone), fl. & fr. (male & female), 9.x.1963, *Leach & Bayliss* 11885 (K; SRGH). M: Inhaca Island, Ponta Ponduine Head, 37 km east of Maputo, fl. (male), 24.ix.1958, *Mogg* 28317 (K; SRGH).
Widespread and naturalised throughout tropical Africa, also in the Arabian Peninsula, Madagascar, and India to the Far East. Locally frequent in open deciduous woodland, often forming thickets in hot dry areas, and often cultivated and naturalised around habitation; 0–1525 m.
Because of the ease with which branch cuttings take root and quickly form dense bushes, this species has been cultivated extensively as a hedging plant. The latex can cause severe eye injury, hence the plant's use in hedging to form an impenetrable barrier. It easily becomes naturalised and eventually develops into a small tree. Usually only male flowers are produced, with female flowers less common. Some with bisexual cyathia also occur, although the female flower often aborts.

61. **Euphorbia gossypina** Pax in Bot. Jahrb. Syst. **19**: 119 (1894). —N.E. Brown in F.T.A. **6**, 1: 553 (1911). —S. Carter in F.T.E.A., Euphorbiaceae, part 2: 474 (1988) pro parte, excl. distr. in Zimbabwe. Type from Tanzania.

Subsp. **mangulensis** S. Carter in Kew Bull. **54**: 959 (1999). Type: Zimbabwe, Makonde Distr., c. 9.5 km southeast of Mhangura (Mangula), fl. & fr. 24.iii.1969, *Leach, Biegel & Pope* 14236 (K, holotype; BR; SRGH).
Euphorbia gossypina sensu Leach in Bothalia **11**: 505 (1975).

Spineless succulent scrambling much branched shrub to 4 m high; branches fleshy, terete, c. 6 mm in diameter, with prominent dark brown leaf scars. Leaves sessile, to c. 30(60) × 6.5 mm, linear-lanceolate, reflexed, soon deciduous. Cymes in 3–8-branched umbels, with simple rays to 2.5 cm long surrounding a central sessile cyathium; umbellate bracts to 15 × 10 mm, ovate; involucral bracts to 8 × 7 mm, ovate, deciduous at the fruiting stage. Cyathia on peduncles to 1.5 mm long, c. 3.5 × 6.5 mm with funnel-shaped involucres; glands 4, or 5–8 on the central cyathium, c. 2.5 × 3 mm, elliptic, yellow; lobes c. 1 × 1.5 mm, subquadrate, denticulate. Male flowers: bracteoles numerous, plumose, conspicuous but scarcely exserted from the involucres; stamens 5 mm long. Female flower: styles c. 3 mm long, joined at the base with spreading deeply bifid apices. Capsule c. 5.5 × 7.5 mm, obtusely lobed, exserted on a recurved pedicel to 7.5 mm long. Seeds to 3.8 × 3 mm, ovoid, smooth, faintly mottled; caruncle c. 1 mm across.

Zimbabwe. N: Makonde Distr., Umbowe Valley, Whindale Farm, Mhangura (Mangula), fl. & fr. 18.ii.1968, *W.B.G. Jacobsen* 3372 (SRGH); Whindale Farm, fr. 14.ii.1968, *Wild* 7686 (SRGH).
Not known elsewhere. Scrambling through shrubs, or pendent on chert cliffs in deciduous bushland; 1160–1200 m.
Subsp. *gossypina*, with bracteoles conspicuously exserted from the involucre, is confined to East Africa. Its variety *coccinea* Pax, with persistent red involucral bracts and green glands, is confined to east of the Rift Valley of southern Kenya and northern Tanzania.

Sect. 14. EUPHORBIA* —S. Carter in F.T.E.A., Euphorbiaceae, part 2: 475 (1988).

Glabrous perennial herbs, shrubs or trees with copious and often strongly caustic latex; monoecious; stems and branches spiny, succulent, cylindrical or longitudinally ridged (angled), sometimes constricted into segments; angles projecting longitudinally, rounded (see TAB. **80**, fig. B2), or ± wing-like and straight or wavy (undulate) (see TAB. **78**, fig. A2), the margins of the angles straight or sinuate (shallowly lobed), with tooth-like projections which are the equivalent of the tubercles of Sections 9 and 10. Leaves usually very small, stipulate, sessile and soon withering (marcescent) or deciduous, or leaves occasionally large, petiolate and more persistent; leaf scars at the apex of spine shields (horny pads). Spine shields crowning the tubercles along the angles, separate and usually decurrent, or sometimes joined to form a continuous horny margin along the angle; each spine shield bearing a pair of spines below the leaf attachment, and with stipules usually modified as prickles (small spines), which are sometimes vestigial. Inflorescence consisting of cyathia in axillary dichasial cymes; cymes usually simple (branching once) with lateral cyathia at right angles to the branch (arranged horizontally), or sometimes parallel (arranged vertically), subsessile or shortly pedunculate, solitary, or 2–5 in horizontal or vertical rows, or in groups of 5 or more, usually developing successively, their point of emergence (flowering eye) situated shortly above the spine shield and sometimes enclosed by the continuous horny margin. Bracts paired, shorter than the cyathia, denticulate on the margins, usually caducous. Central (primary) cyathium often with the involucre containing male flowers only; lateral (secondary) cyathia with bisexual involucres containing male flowers and 1 female flower. Involucres with 5 entire glands and 5 alternating fringed lobes. Male flowers bracteolate, in 5 groups (fascicles) which are usually separated by a fringed membrane; stamens exserted from the involucre. Female flower subsessile or pedicellate, perianth reduced to a rim below the ovary, or sometimes consisting of 3 toothed lobes. Capsule distinctly 3-lobed, occasionally subglobose subsessile or exserted. Seeds subglobose or ovoid, usually minutely verrucose (with warts), without a caruncle.

1. Plants with a definite central stem to 0.5–15 m or more in length and at least 5 cm in diameter; branches angular or winged, up to 4 cm wide or more, rarely less; trees to 30 m tall (see TAB. **78**, fig. B1), or plants with a shrubby habit to 3 m tall · · · · · · · · · · · · · · · 2
- Plants without an obvious central stem; branches angular or ± cylindrical usually 0.5–4.5 cm in diameter, rarely more; dwarf perennials, or shrubs with many stems (see TAB. **80**, fig. A1) · 23

---

* The account of this Section is based with acknowledgement, on the published papers of L.C. Leach.

2. Trees 4–16(30) m high; spine shields always quite separate · · · · · · · · · · · · · · · · · · 3
 – Plants with a shrubby habit 1–3 m high, or trees 4–18 m high; spine shields joined (confluent), to form a horny margin to the angles, at least on older growth · · · · · · · · 6
3. Branchlets (ultimate branches) to 2 cm wide · · · · · · · · · · · · · · · · · · 76. *grandidens*
 – Branchlets 5–17 cm wide · 4
4. Spine shields decurrent, very narrow · 64. *bougheyi*
 – Spine shields c. 5 mm in diameter, rounded to very obtusely triangular · · · · · · · · · · · · 5
5. Branchlets thinly 2–5-winged · 62. *ampliphylla*
 – Branchlets 4-angled or stoutly 4-winged · 63. *ingens*
6. Branches c. 2.5 cm wide, obtusely 3–5-angled (see TAB. **78**, fig. B2), eventually terete; capsule 9 × 25 mm, deeply acutely lobed, exserted on a pedicel c. 8 mm long · · · · · · · 75. *lividiflora*
 – Branches 2–20 cm wide, distinctly winged or angled, not becoming terete, segmented; capsule 2–21 mm in diameter, usually obtusely lobed, subsessile or exserted · · · · · · · · · 7
7. Capsule 6–21 mm in diameter, subsessile on a stout erect pedicel; branches deeply constricted into rounded, ovate, or oblong segments 2.5–20 cm wide, deeply angled · · · 8
 – Capsule 2–8 mm in diameter exserted on a slender recurved pedicel; branches constricted into parallel-sided, rarely ovate segments (83. *memoralis*), 1.5–10 cm wide, shallowly or deeply angled · 15
8. Branch segments 7.5–20 cm wide · 9
 – Branch segments 2.5–7 cm wide · 12
9. Longest spines to 7 cm long · 71. *grandicornis*
 – Longest spines to 1.5 cm long · 10
10. Branch segments oblong, to 32 × 20 cm, wings thin undulate (see TAB. **78**, fig. A2); cymes grouped 1–6 at each flowering eye · 65. *halipedicola*
 – Branch segments mostly ovate-conical; cymes 1–3 in a horizontal line · · · · · · · · · · · · 11
11. Densely branched shrub to 2 m high · 66. *angularis*
 – Tree 2–6 m high, with usually simple arcuate-ascending branches · · · · · · · · · · 67. *cooperi*
12. Shrub to 2 m high; branch segments 3–15 × 3–5 cm, oblong-ovate, 5–7-angled · · · · · · ·
  · 70. *rowlandii*
 – Shrubs or trees 1–5 m high; branch segments 2.5–7 cm wide, rounded-ovate, 3–5-angled
  · 13
13. Shrub to 1 m high; branch segments 2.5–5.25 cm wide, acutely 3-angled · · 72. *williamsonii*
 – Shrubs or trees 3–5(7) m high; branch segments 5–7 cm wide, stoutly 3–5-angled · · · · 14
14. Tree to 5(7) m high; branch segments regularly constricted (hardly separated by narrow sections); 3(4)-angled · 68. *fortissima*
 – Shrub or tree to 3 m high; branch segments rounded, c. 6 × 6 cm, separated by narrow sections to 10 × 2 cm); 3–5(6)-angled · · · · · · · · · · · · · · · 69. *seretii* subsp. *variantissima*
15. Shrub to 1.5(3) m high; branch segments 5–15 × 5 cm, subcircular to ovate, stoutly 4–8-angled (see TAB. **79**, fig. B2) · 83. *memoralis*
 – Shrubs or trees 1–18 m high; branch segments oblong, parallel sided, 3–6-angled · · · · 16
16. Tree to 9–18 m high; branch segments to 10 cm wide, deeply 3–5-angled · · 77. *triangularis*
 – Shrubs or trees 1–8 m high; branch segments no more than 6 cm wide · · · · · · · · · · · 17
17. Cymes 1–3 at each flowering eye, usually in a horizontal line with lateral cyathia arranged vertically · 18
 – Cymes solitary with lateral cyathia arranged horizontally · · · · · · · · · · · · · · · · · · 21
18. Trees to 6–8 m high; branch segments 3–4 cm wide · 19
 – Shrubs, often tree-like, to 3 m high; branch segments 4–6 cm wide · · · · · · · · · · · · · 20
19. Tree to 8 m high (see TAB. **79**, fig. A1); branches mostly 4-angled · · · · · · · · 78. *confinalis*
 – Shrub or tree 2–6 m high; branches mostly 5-angled · · · · · · · · · · · · · · · · · · 79. *keithii*
20. Shrubby tree to c. 2 m high; branch segments to 25 cm long, slightly tapering; spines to 8 mm long · 80. *graniticola*
 – Shrubby tree to 3 m high; branch segments to 15 cm long, sides parallel; spines to 6 mm long · 81. *decliviticola*
21. Plants to 1 m high with a stout trunk to 50 cm high; branches simple · · · · · 82. *mlanjeana*
 – Plants shrub-like, to 3.5 m high; branches rebranching · 22
22. Branches 1–1.5 cm in diameter; leaves to 20 × 6 mm, quickly deciduous · · · · · · · · · · · ·
  · 85. *griseola* subsp. *mashonica*
 – Branches 2–5 cm in diameter; leaves to 10 × 2.5 cm, persistent on young growth · · · · · · ·
  · 84. *persistentifolia*
23. Dwarf perennials; branches constricted into ovate segments 1.5–2 × 1–4.5 cm; capsules sessile · 24

- Shrubs or dwarf perennials; branches not constricted into ovate segments, 0.5–3 cm in diameter · 25
24. Branches to 4.5 cm wide, 3-angled, tightly spiralled; spine shields joined in a horny margin along the angles; longest spines to 2.5 cm long · · · · · · · · · · · · · · · · · · · 73. *tortirama*
- Branches to 3 cm wide, mostly 3-angled; spine shields separate; longest spines to 1.5 mm long · 74. *clavigera*
25. Capsules exserted on a slender recurved pedicel (see TAB. **83**, fig. A3) · · · · · · · · · · · 26
- Capsules sessile or subsessile on a short erect pedicel · 34
26. Branches 1–2.5 cm in diameter; spine shields decurrent to more than halfway to the flowering eye below · 27
- Branches less than 1 cm in diameter; spine shields not decurrent, or decurrent to much less than halfway to the flowering eye below · 30
27. Cyathia greenish-yellow with yellow to reddish glands · 28
- Cyathia entirely dark red · 29
28. Shrub or shrublet c. 30–100 cm high (see TAB. **80**, fig. A1); branch angles shallowly tuberculate · 85. *griseola*
- Shrublet c. 15 cm high; branch angles prominently tuberculate · · · · · · · · · · · 88. *jubata*
29. Shrublet to 40 cm high; branches rarely rebranching · · · · · · · · · · · · · · · 86. *rugosiflora*
- Shrub or shrublet 12–125 cm high, densely branched · · · · · · · · · · · · · · · 87. *richardsiae*
30. Spine shields decurrent to 1 cm · 89. *knuthii*
- Spine shields not decurrent, crowning the tips of the tubercles (see TAB. **81**, fig. B1) · · 31
31. Cymes on peduncles 2–4 cm long produced in a cluster amongst withered branches at the apex of the tuberous root · 90. *decidua*
- Cymes on peduncles 1–5 mm long produced from the branches at the apex of the tuberous root · 32
32. Branches 5–6-angled, 5–7 mm in diameter (see TAB. **81**, fig. B1) · · · · · · · · 91. *fanshawei*
- Branches 3-angled, 3–5 mm in diameter · 33
33. Tubercles on the angles 8–25 mm apart; leaves c. 2 × 1 mm, deciduous · · 92. *mwinilungensis*
- Tubercles on the angles 20–45 mm apart; leaves c. 20 × 5 mm or more, subpersistent · 93. *platyrrhiza*
34. Shrubs 1–2.5 m high · 35
- Shrublets less than 1 m high, usually less than 50 cm · 41
35. Branches 2.5–3 cm in diameter, 4–7-angled but mostly 5-angled · · · · · · · · · · 95. *contorta*
- Branches 1–2 cm in diameter, 4(5)-angled · 36
36. Spines and prickles minute to 2 mm long (see TAB. **83**, fig. B1); shrub to 2.5 m high · 94. *ambroseae*
- Spines 4.5–8 mm long, prickles minute to 2 mm long · 37
37. Branches 12–20 mm in diameter; spines to 4.5 mm long, prickles minute · · · 96. *baylissii*
- Branches to 10 mm in diameter; spines to 5–8 mm long, prickles obvious (see TAB. **82**, fig. B2) · 38
38. Branches greyish-green with conspicuous darker mottling along the angles; shrub branching densely from the base; plants from exposed granite and rocky outcrops, widely distributed · 97. *malevola*
- Branches colouring not as above; shrubs branching from the base sparingly rebranched above; plants from wet areas, beside waterfalls and seasonally inundated woodland · · · 39
39. Branches bright green; spinescence pale grey; angles with tubercles scarcely prominent · 99. *cataractarum*
- Branches and spinescence not as above; angles with tubercles prominent · · · · · · · · · · 40
40. Spines c. 6 mm long, prickles c. 2 mm long, all widely diverging; capsule green with reddish stripes on the sutures · 100. *inundaticola*
- Spines to 8 mm long, ± deflexed, prickles to 1.5 mm long, ± parallel; mature capsule bright red · 101. *luapulana*
41. Small shrubs to 50–90 cm high · 42
- Shrublets usually no more than 30 cm high · 44
42. Shrub to 50 cm high; branches 5–7.5 mm in diameter · · · · · · · · · · · · · · · 98. *dissitispina*
- Shrubs more than 50 cm high; branches 1–15 mm in diameter · · · · · · · · · · · · · · · · · 43
43. Branches blue-green with a conspicuous darker stripe along the angles; involucral glands orange-red · 102. *speciosa*
- Branches uniformly green; involucral glands yellow · · · · · · · · · · · · · · · · · · 103. *perplexa*
44. Spine shields separate, not forming a horny margin along the angles · · · · · · · · · · · · · 45
- Spine shields joined in continuous longitudinal horny ridges along the angles · · · · · · 57

45. Spines 2–17 mm long, obviously longer than the prickles · 46
– Spines and prickles ± equal in length, or the prickles longer than the spines · · · · · · · 53
46. Branches 10–20 mm in diameter; spinescence very sturdy, with spine shields to 12 mm long
    and spines 8–17 mm long · 106. *limpopoana*
– Branches 5–10 mm in diameter; spines shields to c. 8 mm long, spines to 9 mm long, rarely
    to 12 mm long · 47
47. Branches with 3–5 angles spirally arranged · 111. *distinctissima*
– Branches 4-angled or predominantly so · 48
48. Plant a tangled clump to 1 m in diameter; branches rhizomatous, 30–75 cm long · · · · 49
– Plant densely tufted; branches c. 15 cm long, forming cushions c. 50 cm in diameter · · 50
49. Branches spreading, occasionally forming further plantlets; angles with shallow tubercles;
    spines 7–9 mm long, prickles 2 mm long · 104. *dedzana*
– Branches numerous arising densely at and below ground level; producing further plantlets
    to form clumps to 1 m in diameter (see TAB. **83**, fig. C1); tubercles prominent; spines 5–12
    mm long, prickles to 0–1 mm long · 107. *schinzii*
50. Branching irregular or in loose tufts · 51
– Branching in dense tufts forming compact cushion-like plants · · · · · · · · · · · · · · · · · 52
51. Branching irregular; branches ± terete (very obtusely angled); spine shields broadly
    obovate, 2.5–4 × 2–2.5 mm · 108. *venteri*
– Branching tufted forming loose cushion-like plants (see TAB. **84**, fig. A1); branches sharply
    4-angled; spine shields narrow, c. 9 × 1.5 mm · 105. *tholicola*
52. Branches 5–8 mm in diameter; tubercles 6–10 mm apart; spines 2–8 mm long, prickles
    minute · 109. *tortistyla*
– Branches 10–12 mm in diameter; tubercles 10–17 mm apart; spines to 8 mm long, prickles
    0.5–2 mm long · 110. *acervata*
53. Branches 4–5 mm in diameter · 54
– Branches 8–10 mm in diameter · 55
54. Branches up to 25 cm long; spine shields up to 12 mm long, very slender; spines up to 3
    mm long, prickles to 2.5 mm long · 113. *torta*
– Branches up to c. 8 cm long; spine shields up to 6 × 2 mm, obovate; spines and prickles 3–4
    mm long (see TAB. **84**, fig. B3) · 114. *plenispina*
55. Branches 5–7-angled · 115. *whellanii*
– Branches 4-angled or subcylindric · 56
56. Branches obtusely angled; spine shields up to 9 × 2 mm; spines and prickles up to 5 mm
    long · 112. *isacantha*
– Branches subcylindric; spine shields up to 4 × 3 mm; spines and prickles minute up to 3 mm
    long · 116. *debilispina*
57. Branches 4-angled; horny ridges c. 1.5 mm wide · · · · · · · · · · · · · · · · · · · 117. *ramulosa*
– Branches 6–8-angled; horny ridges covering the stem with grooves between · · · · · · · · ·58
58. Spines in pairs, up to 8 mm long, prickles obsolete · · · · · · · · · · · · · · · · · 118. *corniculata*
– Spines single up to 6 mm long, prickles up to 1.5 mm long · · · · · · · · · · · · · 119. *unicornis*

62. **Euphorbia ampliphylla** Pax in Annuario Reale Ist. Bot. Roma **6**: 186 (1897), as
    "*amplophylla*". —Gilbert in Kew Bull. **45**: 196 (1990). —White, Dowsett-Lemaire &
    Chapman, Evergr. For. Fl. Malawi: 253 (2001). Type from Ethiopia.
        Euphorbia winkleri Pax in Bot. Jahrb. Syst. **30**: 342 (1901). —N.E. Brown in F.T.A. **6**, 1: 593
    (1912). —Brenan, Check-list For. Trees Shrubs Tang. Terr.: 214 (1949). Type from
    Tanzania.
        Euphorbia obovalifolia sensu N.E. Brown in F.T.A. **6**, 1: 594 (1912). —sensu Hargreaves,
    Succ. Spurges Malawi: 8 (1987); et sensu auctt. non A. Rich.

    Tree to 10(30) m high; trunk simple to c. 90 cm in diameter; bark rough greyish.
Primary branches ascending, rebranching irregularly and ± densely to form a
spreading rounded crown; branchlets fleshy, 3(4)-angled, 5–17 cm wide, ±
constricted at irregular intervals into oblong segments 15–40 cm long; angles deep
and thinly wing-like, 2–7 cm wide, margins of the angles straight to ± sinuate bearing
tubercles 1.5–3 cm apart. Spine shields rounded, to 5 mm in diameter, eventually
extended to 5 mm above the spines to enclose the flowering eye; spines stout, 1–3
mm long; prickles absent or rudimentary; spine shields and spines becoming corky
and disintegrating. Leaves persistent on young growth, to 15 × 6 cm, obovate, fleshy,
midrib keeled on lower surface, base tapering to a petiole c. 1 cm long. Cymes 1 or
2–3 in a horizontal line, or occasionally 4 crowded together, 1-forked lateral cyathia

vertically arranged; peduncles and cyme branches stout, 3–4 mm long; bracts rounded, 4 × 7 mm. Cyathia c. 4 × 7 mm, with cup-shaped involucres; glands transversely elliptic, c. 2 × 4 mm, margin undulate, not quite touching, yellow; lobes c. 2 × 3.5 mm, transversely elliptic. Male flowers: bracteoles spathulate, plumose; stamens 6 mm long. Female flower: perianth irregularly divided into 5–12 lanceolate toothed lobes to 6 mm long; styles 3.5 mm long, joined at the base, apices thickened, bifid. Capsule to 12 × 16 mm, subglobose, fleshy, red, hardening before dehiscence to become obtusely 3-lobed with woody walls 2–3 mm thick; pedicel stout, 3 mm long. Seeds 4.5 mm in diameter, subglobose, slightly laterally compressed, yellowish-brown, speckled, smooth.

Zambia. N: Chama Distr., Nyika Plateau, Chowo Forest, st. 13.viii.1975, *Pawek* 10031 (K; MAL; MO; SRGH). **Malawi**. N: Rumphi Distr., Nyika Plateau, Nkhonjera Hill, st. 12.vi.1985, *Dowsett-Lemaire* 132 (BR).
Also at high altitudes in Ethiopia, Uganda, Kenya and Tanzania; confined, in the Flora Zambesiaca area to the Nyika Plateau and Matipa Forest (Malawi N). Afromontane rain forest; 2000–2165 m.

63. **Euphorbia ingens** E. Mey. ex Boiss. in de Candolle, Prodr. **15**, 2: 87 (1862). —N.E. Brown in F.C. **5**, 2: 369 (1915). —Burtt Davy, Fl. Pl. Ferns Transvaal: 296 (1932). —Pole Evans (ed.) in Fl. Pl. South Africa **14**: pl. 522 (1934). —White, Dyer & Sloane, Succ. Euphorb. **2**: 925 (1941). —Suessenguth & Merxmüller, [Contrib. Fl. Marandellas Distr.] Proc. & Trans. Rhod. Sci. Ass. **43**: 84 (1951). —Coates-Palgrave, Trees Central Africa: 172 (1957). — Drummond in Kirkia **10**: 253 (1975). —Coates-Palgrave, Trees Southern Africa: 447 (1977). —Court, Succ. Fl. South. Africa: 19 & 25 (1981). —Hargreaves, Succ. Spurges Malawi: 11 (1987); Succ. Botswana: 7 (1990). Type from South Africa.
Euphorbia similis A. Berger, Sukk. Euphorb.: 69 (1907). —N.E. Brown in F.T.A. **6**, 1: 591 (1912); in F.C. **5**, 2: 370 (1915). Type from South Africa.
Euphorbia natalensis A. Berger, Sukk. Euphorb.: 71 (1907). —N.E. Brown in F.C. **5**, 2: 370 (1915), nom. illegit., non Bernh. Type from South Africa.
Euphorbia sp. near quadrialata of Miller, Woody Pl. Bechuanaland Prot.: 44 (1952).
Euphorbia sp. 2 of F. White, F.F.N.R.: 199 (1962).

Massive succulent tree to 4–12(15) m high; trunk stout simple; bark rough fissured grey. Branches persistent from c. 3 m upwards, suberect, rebranching to form eventually a large broadly rounded crown; terminal branchlets fleshy, 4-angled, 6–12 cm wide, square in cross-section to distinctly but stoutly winged with wings to 3 cm wide, usually constricted at irregular intervals into oblong segments to 10–15 cm long or more; margins of the angles straight to sinuate bearing shallow tubercles 1–2 cm apart. Spine shields to 6 × 5 mm, very obtusely triangular, extending 5 mm above to include the flowering eye; spines stout, to c. 5 mm long; prickles flexible, triangular, 1.5 mm long, soon deciduous; spines and spine shields soon becoming corky, rusty-brown and disintegrating. Leaves to 8 × 2 cm and oblanceolate on seedlings and young growth, 3 × 3 mm and deltoid on older growth, and soon deciduous. Cymes crowded towards the apex of the branches, 1–3 at each flowering eye, 1-forked with stout peduncles to 8–20 mm long and cyme branches 5 mm long arranged vertically; bracts to 5 × 6 mm, rounded. Cyathia 5 × 10 mm, with broadly cup-shaped involucres; glands transversely elliptic, 2 × 4 mm, touching, golden-yellow; lobes transversely elliptic, 2.2 × 3 mm. Male flowers many; bracteoles spathulate, plumose; stamens c. 5.5 mm long. Female flower: perianth irregularly divided into 3 or more filiform lobes 2–4 mm long, sometimes with 1 or 2 teeth; styles 3 (rarely 2), 3–3.5 mm long, joined for 1.5 mm, apices thickened, rugulose, bifid. Capsule shortly exserted on a stout pedicel 5 mm long, (2)3-locular, subglobose, 7 × 10 mm, fleshy, green becoming red, hardening immediately before dehiscence to 6 × 9 mm, and very obtusely (2)3-lobed. Seeds subglobose, slightly compressed laterally, to 4 × 3 mm, greyish-brown speckled with paler brown, smooth.

Botswana. N: Xharatshaa, Gomati R., fl. 12.v.1976, *P.A. Smith* 1726 (SRGH). **Zambia**. W: Zambezi Distr., Mombezi–Kabompo confluence, fr. 10.v.1953, *Holmes* 1095 (FHO). S: Mazabuka Distr., Kafue R., fr. vii.1930, *Hutchinson & Gillett* 3585 (K). **Zimbabwe**. N: Guruve Distr., Nyamunyeche Estate, st. 17.xi.1978, *Nyariri* 491 (SRGH). W: Matobo Distr., Mtshelele (Mtsheleli) Valley, fl. 1.v.1952, *Plowes* 1455 (K; SRGH). C: Harare, fl. v.1927, *Eyles* 5857 (SRGH). E: Mutare (Umtali), st. 11.xi.1930, *Fries, Norlindh & Weimarck* 2895 (K). S: Chivi Distr., 3 km south of Runde (Lundi) R. bridge on Masvingo–Beitbridge road, fl. 21.iv.1961, *Leach* 10802 (K;

SRGH). **Malawi**. N: 13 km northwest of Rumphi, st. 6.xi.1977, *Pawek* 13218 (BR; K; MAL; SRGH; WAG). C: Nkhotakota Game Reserve, Chipata Hill, fr. 9.viii.1981, *Chapman* 5846 (K; SRGH). S: Zomba Distr., Chaone Island, Lake Chilwa, fr. 1.vi.1970, *Brummitt & Williams* 11218 (K; MAL; SRGH). **Mozambique**. N: Mueda Distr., 25 km from Chomba on Negomano road, fl. 13.iv.1964, *Torre & Paiva* 11883 (LISC; SRGH). T: Cahora Bassa Distr., Songo, fl. 3.iv.1972, *Macêdo* 5148 (LISC; SRGH). GI: 13 km from Mabote on L. Banamana road, fl. 8.ix.1973, *Correia & Marques* 3277 (WAG). M: 19.5 km from Matutuíne (Bela Vista) on road to Porto Henrique, fl. 23.iv.1964, *Balsinhas* 702 (LISC).

Also in eastern Angola, and eastern parts of South Africa and Swaziland. Common from low to higher altitudes, usually solitary, in dry mopane and wooded grassland, often on rocky outcrops; 10–1500(1600) m.

Very similar to *E. candelabrum* Kotschy from east and northeast Africa, and may be conspecific (according to L.C. Leach). However, branches of the southern trees appear usually to be more distinctly and shortly segmented, the tubercle teeth along the angles are mostly further apart, and cymes are generally fewer at each flowering eye. In view of the long established name of *E. ingens* for the southern plants, and the controversy over the validation of Kotschy's name (see S. Carter in F.T.E.A. loc. cit.: 486 for details) for the northern plants, it seems most practical at present to treat populations from the two regions as two species.

64. **Euphorbia bougheyi** L.C. Leach in J. S. African Bot. **30**: 9, fig. & photos on pl. V (1964). — Court, Succ. Fl. South. Africa: 27 (1981). Type: Mozambique, Beira, Macúti Beach, fl. vii.1963, *Boughey* in *GHS* 12206 (SRGH, holotype; CAH; COI; G; K; LISC; LMU; PRE).

Succulent tree to 7 m high; trunk cylindric, fairly slender, spiny above, with up to 9 angles and a crown of verticillate, spreading branches, curving upwards. Branches 2–5-winged, constricted into variably shaped segments 2.5–5 cm wide; wings very thin, margins prominently to obscurely crenate, usually distinctly wavy, with crenations 10–25 mm apart; ultimate branchlets often 2-winged, produced in whorls. Spine shields extremely narrow, usually separate, sometimes extended above to enclose the flowering eye 3–5 mm above the spine pairs; spines to 7.5 mm long; prickles minute. Leaves rudimentary, ovate-acute, recurved, withered remains usually persistent. Cymes solitary; peduncles and cyme branches very stout, c. 2.5 × 4 mm; bracts c. 4 × 7 mm, broadly ovate, usually split to the base. Cyathia to 4.5 × 10 mm, with broadly funnel-shaped involucres; glands 5–5.5 mm wide, transversely narrowly elliptic, spreading, rugulose, yellow; lobes broadly elliptic, irregularly dentate. Male flowers c. 60: fascicular bracts c. 4 mm long, laciniate; stamens c. 5.5 mm long. Female flower: perianth 3-lobed, lobes 1.5 mm long, sometimes with 1–2 teeth; styles to 3 mm long, free almost to the base, spreading recurved. Capsule 8.5–9.5 × 19–23 mm, deeply lobed, exserted on a stout pedicel c. 7 mm long. Seed c. 3.75 mm in diameter, subglobose.

Mozambique. Z: Maganja da Costa Distr., Floresta de Gobene, 45 km from Olinga (Vila da Maganja), st. 14.ii.1966, *Torre & Correia* 14612 (LISC). MS: Machanga Distr., between Divinhe and Nova Mambone, fr. 2.ix.1961, *Leach* 11254 (CAH; COI; G; K; LISC; LM; PRE; SRGH). GI: Vilankulo (Vilanculos) turn-off on Maxixe–Mambone road, fr. 6.x.1963, *Leach & Bayliss* 11852 (G; K; LISC; LMJ; PRE; SRGH).

Known only from swampy coastal areas to the north and south of Beira. Dense woodland and thickets in low-lying grassland; 0–120 m.

Closely related to *E. halipedicola* and *E. nyikae* Pax (East Africa), but distinguished by its very thin irregularly shaped branch segments.

65. **Euphorbia halipedicola** L.C. Leach in J. S. African Bot. **36**: 42, fig. & photos on pages 46 & 47 (1970). —Drummond in Kirkia **10**: 253 (1975). —Coates-Palgrave, Trees Southern Africa: 447 (1977). —Court, Succ. Fl. South. Africa: 25 & 27 (1981). —Leach in Fl. Pl. Africa **49**: pl. 1928 (1986). —Hargreaves, Succ. Spurges Malawi: 23 (1987). TAB. **78**, fig. A. Type: Mozambique, near Lake Gambué, 22.vi.1961, *Leach & Wild* 11130 (SRGH, holotype; K; LISC; PRE).

Euphorbia angularis sensu N.E. Brown in F.T.A. **6**, 1: 585 (1912), pro parte as to *Kirk* s.n. (Lupata), non Klotzsch.

Succulent shrub or small tree to 4–5(10) m high; trunk stout, with arcuate-ascending branches. Branches sparingly rebranched, 3–4-winged, deeply constricted into segments; segments to 32(40) × 20 cm, usually oblong, or ± broadly deltate or ovate; wings broad, to 2 mm thick, wings conspicuously undulate, with tubercles 1.5–3 cm apart along the margins. Spine shields joined to form a continuous horny

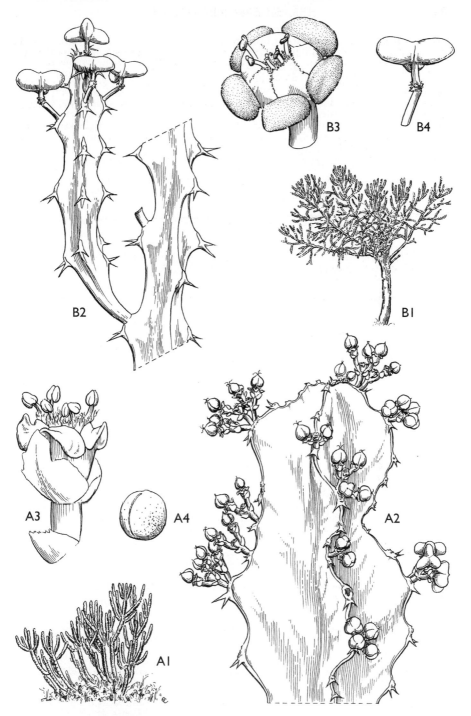

Tab. 78. A. —EUPHORBIA HALIPEDICOLA. A1, habit (× c. $^1/_{120}$), from a photo; A2, fruiting branch (× $^2/_3$); A3, cyathium (× 4); A4, seed (× 4), A2–A4 from *Leach, Cannell & Tinley* 14253. B. —EUPHORBIA LIVIDIFLORA. B1, habit (× c. $^1/_{100}$), from a photo; B2, portion of fruiting branch (× $^2/_3$); B3, cyathium (× 4); B4, capsule (× 1), B2–B4 from *Leach & Wild* 11129. Drawn by Eleanor Catherine.

margin to the wings; spines stout, spreading, to 15 mm long, longest at the widest part of the wings of the branch segments. Cymes 1–6 and randomly arranged at each flowering eye, simple, lateral cyathia vertically arranged; peduncles and cyme branches 7–12.5 mm long, c. 5 mm in diameter; bracts to 4–5.5 × 5.5–7 mm, broadly ovate. Cyathia 6 × 8–10 mm, broadly funnel-shaped; glands c. 3.25 × 4–5 mm, transversely broadly elliptic, outer margin ± undulate; lobes 2.5 mm long, obovate, keeled, margin fimbriate. Male flowers c. 50: bracteoles laciniate; stamens c. 6 mm long. Female flower: perianth 3-lobed, 6 mm in diameter; ovary subglobose; styles c. 4.5 mm long, united for 2 mm, apices shortly bifid, rugulose. Capsule deeply lobed, c. 8 × 18–21 mm, base truncate, exserted on a pedicel to 8 mm long, 3.5 mm in diameter. Seeds c. 3.5 mm in diameter, subglobose, buff mottled brownish.

Zimbabwe. E: Chiredzi Distr., 14 km east of Rimbi Township, Save Valley, fr. 11.viii.1971, *Percy-Lancaster* 3 (SRGH). **Mozambique**. N: Mossuril Bay, between Lunga and Lumbo, fl. 11.viii.1964, *Leach* 12371 (K; SRGH). Z: 3 km north of Mopeia (Mopeia Velha), st. 7.xii.1971, *Müller & Pope* 1933 (LISC; SRGH). MS: Parque Nacional da Gorongosa (Parc Nacional de Caça), fl. & fr. 8.vii.1969, *Leach, Cannell & Tinley* 14253 (K; SRGH). GI: 16 km south of Nova Mambone, fl. (no date), *Ambrose* s.n., cult. in *Leach* 11728 (K; SRGH).

Also in southeastern Tanzania northwards to near Dar es Salaam. In, or at the margins of thickets and small patches of woodland, mostly on the coastal plain and usually in areas of seasonally flooded grassland, often in association with *E. lividiflora*, *Aloe marlothii* and *Hyphaene sp.*; 10–500 m.

Reported by Hargreaves (Succ. Spurges Malawi: 23 (1987)) to occur also at the southern tip of Malawi, Sorjin, *Hargreaves* 1123.

66. **Euphorbia angularis** Klotzsch in Peters, Naturw. Reise Mossamb. **6**: 92 (1861). —N.E. Brown in F.T.A. **6**, 1: 584 (1912) pro parte. —White, Dyer & Sloane, Succ. Euphorb. **2**: 881 (1941) pro parte. —Leach in J. S. African Bot. **36**: 17, photos on pages 20 & 22 (1970). —Court, Succ. Fl. South. Africa: 16 (1981). Type: Mozambique, Goa Island, *Peters* s.n. (B†); ibid., fl. & fr. 10.viii.1964, *Leach* 12361 (SRGH, neotype; K; LISC; MO; PRE; ZSS).

Euphorbia abyssinica var. *mozambicensis* Boiss. in de Candolle, Prodr. **15**, 2: 84 (1862). Type as above.

Succulent shrub up to c. 2 m high, densely branched and rebranched. Main stem much reduced. Branches spreading, secondary branches and branchlets mostly erect, conspicuously segmented; segments to 15 × 12 cm, ovate to oblong, 3(4)-winged or -angled from a stout axis; the angles sinuate with tubercles 10–12 mm apart along the margins. Spine shields joined to form a continuous horny margin 2–5 mm wide; spines spreading, up to 10 mm long, much reduced at the constrictions of the branches; prickles minute, usually hooked; secondary spines usually flanking the flowering eyes shortly above the spine-pairs, straight or hooked, c. 3 mm long, also with further small spines and prickles. Leaves c. 1.5 mm long, broadly ovate-acute, caducous. Cymes 1–3 in a horizontal line, lateral cyathia vertically arranged, often 2–3-forked producing crowded cyathia; peduncles and cyme branches stout, 5–6 mm in diameter; bracts semi-circular, c. 3 × 8 mm, usually splitting. Cyathia c. 4 × 9 mm, with cup-shaped involucres; glands spreading, transversely elliptic, 2.25–3 × 4.5–7 mm, yellow; lobes subquadrate, fimbriate, 2 mm wide. Male flowers 35–40: fascicular bracts lacerate, fimbriate; bracteoles filiform; stamens to 6.25 mm. Female flower: perianth c. 6 mm in diameter, ± triangular; styles stout, 3.5–5 mm long, shortly united up to two-thirds, apices spreading-recurved, shortly bifid. Capsule obtusely lobed, c. 8 × 14–17.5 mm, shortly exserted, pedicel stout, slightly curved, to 7 mm long. Seeds subglobose, c. 3.5 mm in diameter, ± smooth, buff tinged pale purplish, minutely blackish spotted.

Mozambique. N: Mossuril Distr., Goa Island, st. 19.v.1961, *Leach & Rutherford-Smith* 10932 (K; PRE; SRGH); Ilha de Goa, [fruits slightly larger than neotype], fr. 4.viii.1947, *Gomes e Sousa* 3514 (K).

Known only from the tiny coral islet of Goa Island off the coast of Mozambique. On coral rocks; c. 5 m.

Previously confused with *E. cooperi* but distinguished by its short, irregularly 3-angled main stem and densely branched shrubby habit with very variable mostly 3-angled branch-segments, and by its larger more deeply lobed capsule and larger seeds. On Goa Island it is the only species of spiny *Euphorbia*. All known references of its occurrence elsewhere are erroneous and can usually be referred to *E. cooperi*, *E. halipedicola* or *E. ingens*.

67. **Euphorbia cooperi** N.E. Br. ex A. Berger, Sukk. Euphorb.: 83, excl. fig. 21 (1907). —N.E. Brown in F.C. **5**, 2: 367 (1915). —Pole Evans (ed.) in Fl. Pl. South Africa **4**: pl. 157 (1924). —Burtt Davy, Fl. Pl. Ferns Transvaal: 296 (1932). —White, Dyer & Sloane, Succ. Euphorb. **2**: 872 (1941). —Leach in J. S. African Bot. **36**: 27, photos on page 29 (1970). —Drummond in Kirkia **10**: 253 (1975). —Coates-Palgrave, Trees Southern Africa: 441 (1977). —Court, Succ. Fl. South. Africa: 16 & 25 (1981). —Hargreaves, Succ. Botswana: 8 (1990). —S. Carter in F.T.E.A., Euphorbiaceae, part 2: 490 (1988). Type from South Africa (Kwa-Zulu Natal).
  Euphorbia angularis sensu auctt. non Klotzsch.
  Euphorbia grandidens sensu Burtt Davy in Transvaal Agric. J. **4**: t. 38 (1905). —sensu Eyles in Trans. Roy. Soc. South Africa **5**: 399 (1916), non Haw.

Flat-topped candelabriform succulent tree, 2–6(9) m high, shrub-like to c. 2 m high in younger plants; trunk stout cylindric, scarred from fallen branches; branches arcuate-ascending, shortly exceeding the trunk apex, usually simple or occasionally branched near the apices, deeply and variably segmented; segments 10–50 × 5–20 cm, (2)3–7(8)-winged; the angles shallowly sinuate-undulate, with tubercles 8–25 mm apart along the margins. Spine shields joined to form a continuous horny margin to the wings; spines to 10 mm long; prickles vestigial. Leaves c. 1.5 × 1.5 mm, deltate. Cymes 1–3, horizontally arranged, with cyathia vertically arranged, subsessile. Cyathia c. 4.5 × 8 mm, with cup-shaped involucres; glands c. 1.5 × 4 mm, transversely oblong, touching, golden-yellow; lobes c. 1.5 × 2 mm, subquadrate. Male flowers: bracteoles deeply laciniate, plumose. Female flower: perianth shallowly lobed; styles variable, from c. 2 mm and free almost to the base, to 5.5 mm and united for two-thirds, apices bifid. Capsule variable in size and degree of lobing, 6–10 × 10–13.5 mm; pedicel stout, 4–10 mm long. Seeds 2.8–3.5 mm in diameter, subglobose, pale greyish-brown, speckled, smooth.

1. Shrubs or small trees to 4 m high; winged angles 2–4, thin; segments usually wider than long · ii) var. *calidicola*
- Shrubs or trees to 9 m high; winged angles 3–8, stout; segments circular or usually longer than broad · 2
2. Branch segments mostly conic-ovate; capsules to 7.5 × 13.5 mm on a pedicel to 10 mm long · i) var. *cooperi*
- Branch segments mostly subcircular; capsules to 6 × 10 mm on a pedicel 3–5 mm long · iii) var. *ussanguensis*

i) Var. **cooperi**

Plants mostly less than 6 m high. Branches mostly simple, occasionally with branchlets near the apices, (3)4(6)-winged; segments variably shaped, mostly conic ovate or sub-elliptic, c. 1.5 times as long as broad. Spine shields forming ± even, continuous, horny margins 5–6 mm wide; spines stout, to 10 mm long. Capsule variable, generally ± triangular and c. 10 mm high, to ± lobed and up to 13.5 × 7.5 mm, exserted on a slightly curved pedicel to 10 mm long.

Botswana. SE: Moeng, fl. 26.vii.1980, *Woollard* 784 (SRGH). **Zimbabwe.** N: 3 km north of Mutoko (Mtoko), fl. 24.vii.1959, *Leach* 9244 (BR; K; PRE; SRGH). W: Umzingwane Distr., Diana's Pool, Matopos, fr. 12.xi.1955, *Plowes* 1875 (K; LISC; SRGH). C: Battlefields, fl. 31.viii.1956, *Whellan* 1121 (SRGH). E: Hot Springs, fl. & fr. 25.ix.1965, *Leach & Müller* 13137 (K; LISC; PRE; SRGH; ZSS). S: Runde (Lundi) R. bridge on Beitbridge road, fl. 6.v.1963, *Leach* 11647 (K; LISC; M; MO; PRE; SRGH; ZSS). **Mozambique.** N: 40 km east of Malema (Entre Rios), fl. 16.xii.1967, *Torre & Correia* 16539 (LISC). GI: Dumela, fl. 11.vii.1964, *Leach & Mockford* 12300 (BM; BR; EA; G; K; LISC; MO; PRE; SRGH). M: Namaacha, fl. 2.v.1964, *Balsinhas* 717 (LISC).
This is the most widespread variety, and is common almost throughout Zimbabwe. It extends marginally into Botswana, and into southern Mozambique, South Africa (Northern Province, northern KwaZulu-Natal) and Swaziland. At low to medium altitudes, on rocky hillsides and granite kopjes, usually forming colonies; 200–1450 m.

ii) Var. **calidicola** L.C. Leach in J. S. African Bot. **36**: 34, photos on page 36 (1970). —Drummond in Kirkia **10**: 253 (1975). —Court, Succ. Fl. South. Africa: 25 (1981). —Hargreaves, Succ. Spurges Malawi: 21 (1987). Type: Zimbabwe, Deka/Shashachunda R. junction, fr. 24.viii.1968, *Leach & Cannell* 14103 (SRGH, holotype; K; LISC).
  Euphorbia angularis sensu Boughey in J. S. African Bot. **30**: 162 (1964), non Klotzsch.
  Euphorbia sp. 1 of F. White, F.F.N.R.: 199 (1962).

Plants generally less than 4 m high. Branch segments (2)3–4-winged, usually as wide as long, or wider than long; wings very thin, c. 3 mm thick. Spinescence very variable, with spines occasionally to 4.5 cm long, but usually much reduced, sometimes almost obsolete. Capsule exserted on a pedicel to 10 mm long.

Zambia. W: Luano Valley, fl. 12.vi.1928, *Burr* 4 (K). E: Petauke Distr., east bank of Luangwa R., fr. 5.ix.1947, *Greenway & Brenan* 8057 (FHO; K; PRE). S: Lusitu, hillsides north of Kariba, fl. 22.v.1961, *Fanshawe* 6605 (K; NDO). **Zimbabwe**. N: Binga Distr., between Sebungwe Drift and Binga, fl. 16.v.1955, *Plowes* 1849 (K; PRE; SRGH). W: Hwange Distr., 3 km south of Inyantue R., st. 29.iii.1963, *Leach* 11611 (BR; EA; K; LISC; PRE; SRGH). **Malawi**. N: Nkhata Bay Distr., Likoma Island, fl. 27.viii.1984, *Salubeni, Seyani & Nachamba* 3904 (K; MAL). S: Blantyre Distr., Mpatamanga Gorge, fl. 16.viii.1960, *Leach* 10458 (B; BR; K; PRE; SRGH). **Mozambique**. T: Changara Distr., Mazowe/Luenya (Luenha) R. junction, fl. 6.v.1960, *Leach* 9935 (BM; BR; K; LISC; MO; PRE; SRGH; ZSS). MS: Tambara Distr., between Nhacolo (Tambara) and Belo, fl. 14.vii.1969, *Leach & Cannell* 14330 (BM; K; LISC; LMA; SRGH).

This variety occurs at low to medium altitudes, and is associated mainly with the Zambezi R. (the stretch along Lake Kariba to the Shire R.) and its larger tributaries (including the Luangwa and Shire Rivers); also on Lake Malawi. Riverine thickets and woodland, on rocky slopes; 200–830(1200) m.

iii) Var. **ussanguensis** (N.E. Br.) L.C. Leach in J. S. African Bot. **36**: 31, photos on page 33 (1970). —Court, Succ. Fl. South. Africa: 16 (1981). —Hargreaves, Succ. Spurges Malawi: 22 (1987). —S. Carter in F.T.E.A., Euphorbiaceae, part 2: 491 (1988). Type from Tanzania.
  Euphorbia ussanguensis N.E. Br. in F.T.A. **6**, 1: 587 (1912). —Brenan, Check-list For. Trees Shrubs Tang. Terr.: 214 (1949). —Fanshawe, Check List Woody Pl. Zambia Showing Distrib.: 23 (1973). Type from Tanzania.

Plants usually taller, to 9 m, more freely branched and rebranched. Branch segments ± circular, (3)4–6(8)-angled; angles stoutly winged. Spine shields forming horny margins 3–10 mm wide; spines 3–10 mm long. Capsule distinctly 3-lobed, 6 × 10 mm, scarcely exserted on a straight, stout pedicel 3–5 mm long. Seeds c. 3 mm in diameter.

Zambia. N: Mansa (Fort Rosebery), fl. 11.v.1964, *Fanshawe* 8621 (K; FHO; NDO; SRGH). C: Serenje Distr., c. 51 km north of Kanona, fl. 14.vi.1960, *Leach & Brunton* 10043 (K; LISC; NDO; PRE; SRGH). E: Lundazi, fl. 16.viii.1965, *Fanshawe* 9275 (NDO; SRGH). **Malawi**. N: Rumphi, fl. 17.viii.1972, *Brummitt & Patel* 12902 (K). C: Nkhotakota Distr., Chipata Mt., st. 9.viii.1981, *Chapman* 5845 (K; SRGH).

Also in southern Tanzania. Rocky hillsides and wooded gorges; 1050–1600 m.

68. **Euporbia fortissima** L.C. Leach in J. S. African Bot. **30**: 209, fig. & pl. XXVI (1964). —Fanshawe, Check List Woody Pl. Zambia Showing Distrib.: 23 (1973). —Drummond in Kirkia **10**: 253 (1975). —Coates-Palgrave, Trees Southern Africa: 445 (1977). —Court, Succ. Fl. South. Africa: 25 (1981). Type: Zimbabwe, Hwange Distr., Shashachunga River, c. 11 km north of Hwange, fl. 30.iii.1963, *Leach* 11616 (SRGH, holotype; BM; COI; G; K; LISC; MO; PRE).
  Euphorbia cooperi sensu White, Dyer & Sloane, Succ. Euphorb. **2**: 880, fig. 1003 (1941).

Candelabriform succulent tree to 5(7) m high; trunk stout, to 22.5(30) cm in diameter; crown flat-topped. Branches usually whorled, spreading and curving upwards, usually exceeding the trunk apex, 3(4)-angled, regularly constricted; segments 6–8 × 5–7 cm, ovate to circular; angles stoutly wing-like, with shallow tubercles c. 1 cm apart along the margins. Spine shields forming a continuous horny margin to 8 mm wide along the angles; spines stout, 2–7 mm long, diverging; prickles rudimentary; rudimentary secondary spines flanking the flowering eyes on young growth. Leaves minute, quickly deciduous. Cymes 1–3 from each flowering eye, lateral cyathia vertically arranged; peduncles 7–10 mm long, cyme branches 5–8 mm long, stout to 4.5 mm in diameter; bracts c. 7.5 mm wide, broadly elliptic, obtusely keeled, usually ± split. Cyathia 4–5 × 10–11 mm, with cup-shaped involucres; glands 2.5 × 5 mm, reniform, spreading, closely touching, yellow; lobes c. 2 × 3 mm, subquadrate. Male flowers c. 60: bracteoles fimbriate; stamens 6 mm long. Female flower: perianth irregularly dentately 3-lobed; styles spreading, 2.25 mm long, joined up to two-thirds. Capsule 9–18 × 15–24 mm, obtusely lobed, truncate at the base and apex, exserted on a stout, curved pedicel to 13 mm long. Seed 4 × 3.5 mm, subglobose, olive-brown, mottled with a waxy coating.

Zambia. S: Livingstone Distr., Candelabra Pool near 5th gorge below Victoria Falls, fl. 7.iii.1964, *Leach* 12121 (K; LISC; PRE; SRGH). **Zimbabwe**. N: Hwange Distr., c. 27 km west of Hwange (Wankie), near Nashome, fl. & fr. 31.ii.1963, *Leach* 11626 (K; PRE).

Known only from the gorges below Victoria Falls and the area between Hwange and Victoria Falls. On stony slopes in *Mopane–Commiphora–Cobretum* woodland; 500–600 m.

Most closely related to *E. cooperi* var. *cooperi*, with stoutly wing-angled segmented branches, continuous horny margins on the angles, and strong spinescence, but with considerably larger cyathia, capsules and seeds.

69. **Euphorbia seretii** De Wild. in Ann. Mus. Congo, Bot., Sér. V, [Études Fl. Bas-Moyen-Congo] **2**: 290 (1908). —N.E. Brown in F.T.A. **6**, 1: 603 (1912). —Court, Succ. Fl. South. Africa: 29 (1981). Type from Dem. Rep. Congo.

Subsp. **variantissima** L.C. Leach in J. S. African Bot. **35**: 15, fig. & photos on page 14 (1969). Type: Zambia, Kabompo Gorge, above the rapids, fl. 25.x.1966, *Leach & Williamson* 13551 (SRGH, holotype; B; BM; BR; G; K; LISC; LUA; LUAI; NDO; PRE; ZSS).

Euphorbia seretii sensu Fanshawe, Check List Woody Pl. Zambia Showing Distrib.: 23 (1973).

Succulent tree to 3 m high, branching from the base, or with a stout trunk to 1.3 m high and 22.5 cm thick. Branches usually simple, to 1.6 m long, curving upwards, 3–5(6)-angled, constricted into segments usually subcircular in outline; segments c. 6 × 6 cm, separated by narrow sections up to 2 cm wide and 10 cm long; angles with shallow tubercles c. 1 cm apart along the margins. Spine shields forming a continuous horny margin along the angles, c. 4 mm wide; spines 1.5–8 mm long; prickles fleshy, minute; minute secondary prickles often flanking the flowering eyes; the whole spinescence rapidly degenerating to become black and corky. Leaves fleshy, vestigial, quickly deciduous. Cymes 1–5 in a horizontal line at each flowering eye, simple, with lateral cyathia vertically arranged; peduncles and cyme branches 2–9 mm long, c. 4 mm in diameter; bracts 2–3 × 3.5–5 mm, triangular, often split. Cyathia 3–3.5 × 5–7.5 mm, with cup-shaped involucres; glands 2.5–3.5 mm wide, transversely oblong, touching, yellow; lobes c. 1.5 mm wide, subquadrate, denticulate; primary cyathium usually bisexual. Male flowers c. 40: bracteoles filiform-fimbriate; stamens c. 4.75–6 mm long, well exserted. Female flower: perianth reduced to a rim; styles to 3 mm long, joined to halfway, with spreading, capitate apices. Capsule c. 4.5 × 9.5 mm, deeply lobed, truncate at the base; pedicel stout, exserted, to 7 mm long. Seeds c. 3 × 2.6 mm, subglobose, brownish, smooth or obscurely rugulose.

Zambia. W: Kabompo Gorge, st. 12.vi.1974, *Chisumpa* 176 (K; NDO; SRGH). C: Mkushi Boma, st. vi.1966, *G. Williamson*, cult. Harare in *Leach* 13537 (BM).

Known only from these specimens. Among rocks in *Brachystegia* woodland and in the gorge; 1000–1200 m.

Subsp. *seretii* is a smaller shrub, with more regular segmentation of the branches, and is restricted to northeastern Dem. Rep. Congo.

70. **Euphorbia rowlandii** R.A. Dyer in Bothalia **7**: 28 (1958). —Court, Succ. Fl. South. Africa: 14 & 25 (1981); in Mem. Bot. Survey S. Africa **62**: 472 (1993). Type from South Africa.

Densely branched succulent shrub-like plant to c. 2 m high. Stem much reduced. Branches widely spreading, becoming erect, rebranching from near the base, 5–7-angled, constricted into segments ovate to ± circular in outline; segments up to 15 cm long and 3–5 cm wide; angles sinuate with shallow tubercles 6–10 mm apart along the margins. Spine shields forming a continuous horny margin c. 2 mm wide along the angles; spines 5–13 mm long; prickles rudimentary or absent. Cymes solitary, simple with lateral cyathia vertically arranged; peduncle and cyme branches c. 2 mm long; bracts 1 × 2.5 mm, deciduous. Cyathia with involucres 3 × 4.5–5 mm; glands 2.5–3 mm wide, transversely oblong, touching, yellow; lobes c. 1.5 × 1.5 mm, subquadrate, fimbriate. Male flowers: bracteoles c. 3 mm long, plumose; stamens 4 mm long. Female flower: styles 1.5 mm long, united for one-third. Capsule c. 9 mm in diameter, acutely lobed, pedicel 0.5–1 mm long. Seeds c. 2.5 mm in diameter, subglobose, brown.

Zimbabwe. S: Chiredzi Distr., Pesu R. Gorge, c. 23 km northwest of Pafuri, north of Limpopo R., st. 10.vii.1964, *Leach & Mockford* 12286 (K; PRE; SRGH).
Also in South Africa (Northern Province, northwest of Punda Maria). Hot dry low altitudes, on cliffs and sandstone ridges; 300 m.

71. **Euphorbia grandicornis** Goebel ex N.E. Br. in Hooker's Icon. Pl. **26**: t. 2531 & 2532 (1897). —Berger, Sukk. Euphorb.: 52 (1907). —N.E. Brown in F.C. **5**, 2: 367 (1915) pro parte excl. *E. breviarticulata*. —R.A. Dyer in Fl. Pl. South Africa **17**: t. 642 (1937). —White, Dyer & Sloane, Succ. Euphorb. **2**: 861 (1941). —Court, Succ. Fl. South. Africa: 15 (1981). Type from South Africa (KwaZulu-Natal).

Succulent shrub 50 cm to 2 m high, branching from the base. Branches ascending or occasionally procumbent, simple or sometimes rebranching near the apices, 2–4-angled, deeply constricted into broad segments obovate in outline; segments 5–12 × 5–15 cm; angles winged, with wings 3–7 cm broad and up to 5 mm thick, the margins irregularly sinuate with tubercles 1–2.5 cm apart along the margins. Spine shields joined in a continuous horny margin; spines 1.5–7 cm long, but reduced to c. 3 mm long at the constrictions of the branches; prickles minute; secondary spines often flanking the flowering eyes, 1–2 mm long. Leaves 1 × 1.5 mm, deciduous. Cymes 1–3 in a horizontal line at each flowering eye, simple, with lateral cyathia vertically arranged; peduncles and cyme branches 3–5 mm long; bracts 2 × 2.5 mm, scale-like. Cyathia with funnel-shaped involucres 4 × 6–8 mm in diameter; glands transversely oblong, 4 mm wide, touching, yellow; lobes subquadrate, fimbriate. Male flowers: bracteoles laciniate-fimbriate, c. 3.5 mm long; stamens c. 5.5. mm long. Female flower: perianth 3-lobed, with lobes c. 1 mm long; styles 4 mm long, united to halfway, apices spreading, capitate. Capsule c. 6 × 13 mm, obtusely lobed, truncate at the base, red when mature, subsessile. Seed c. 2.5 × 2 mm, subglobose, buff with brown speckles, smooth.

Branches 3–4-angled · subsp. *grandicornis*
Branches 2–3-angled; angles wing-like · subsp. *sejuncta*

Subsp. **grandicornis**

Shrub to 2 m high; branches 3–4-angled; male flowers c. 25.

Mozambique. MS: Machaze Distr., Maringa (Maringua) Village, 8 km north of Save (Sabi) R., st. 23.vi.1950, *Chase* 2340 (BM). GI: Guijá Distr., Caniçado, 9 km from Lagoa Nova, fl. 22.vii.1969, *Correia & Marques* 1015 (WAG). M: 16 km south of Boane, fr. 22.vii.1961, *Leach* 11207 (EA; K; PRE; SRGH).
Also in Swaziland and South Africa (KwaZulu-Natal). Dense dry mixed woodland; 40–190 m.

Subsp. **sejuncta** L.C. Leach in J. S. African Bot. **36**: 39 (1970). Type: Mozambique, c. 9.5 km east of Nampula, fl. 23.vii.1962, *Leach & Schelpe* 11437 (SRGH, holotype; K; LISC; MO; PRE).

Shrub to 60(100) cm high; branches sometimes procumbent, 2–3-angled, angles wing-like. Male flowers more numerous, c. 50.

Mozambique. N: Murrupula Distr., Serra de Chinga, 30 km east of Ribáuè, fr. 10.x.1968, *Macêdo* 3704 (LMA).
Known only from the vicinity of the type locality. Slopes of granite outcrops; c. 400–700 m.

72. **Euphorbia williamsonii** L.C. Leach in J. S. African Bot. **35**: 19, fig. & photos on page 20 (1969). —Fanshawe, Check List Woody Pl. Zambia Showing Distrib.: 23 (1973). —Court, Succ. Fl. South. Africa: 29 (1981). Type: Zambia, Kawambwa Distr., Ntumbachushi Falls, fl. & fr. 22.xii.1967, *Williamson & Simon* 729 (SRGH, holotype; PRE).

Succulent shrub up to 1 m high, branching from a partially exposed tuberous root. Branches erect, rarely rebranching, acutely 3-angled, slightly constricted into variably shaped segments; segments 2.5–5 × 2.5–5.25 cm, ± obovate to subcircular; angles crenulate with tubercles to 18 mm apart along the margins, or further apart just above the constrictions. Spine shields ± obovate, forming a ± continuous horny margin along the angles, 1–2.5 mm wide; spines to 1 cm long, diverging; prickles rudimentary; secondary rudimentary spines often flanking the flowering eye. Leaves c. 2 × 1.5 mm, ovate, caducous. Cymes solitary, simple or reduced to single cythia; peduncle 1.5–5.5 × 5.75 mm; bracts c. 2.5 × 5.5 mm, semicircular, blackish-red. Cyathia to 3 × 5–7 mm, with funnel-shaped involucres, bright red; glands 5(6), 1.5 × 2.5–4 mm, transversely elliptic, yellowish with reddish margin; lobes 1.5 × 2–2.5 mm, transversely oblong, fimbriate, dark red.   Male flowers c. 60: bracteoles numerous, fimbriate; stamens c. 5.5 mm long. Female flower: styles stout, 2.5–3 mm long, spreading recurved, united at the base, stigmas capitate. Capsule to 5.5 × 9.5 mm, obtusely lobed, subsessile, shiny bright red. Seeds c. 3.5 × 3 mm, subglobose, smooth, brown.

Zambia. N: Ntumbachushi Falls, c. 16 km west of Kawambwa, fl. & fr. 13.xi.1957, *Fanshawe* 4080 (K; NDO); Ntumbachushi Falls, *A.W. Dock* s.n., cult. Nelspruit, fl. 6.iii.1968, in *Leach* 12801 (K).
Known only from the type locality. In fibrous mats on rocky quartzite outcrop; 1150 m.

73. **Euphorbia tortirama** R.A. Dyer in Fl. Pl. South Africa **17**: t. 644 (1937). —White, Dyer & Sloane, Succ. Euphorb. **2**: 771 (1941). —Court, Succ. Fl. South. Africa: 13 (1981). Type from South Africa (Northern Province).

Caespitose dwarf spiny succulent perennial, with a large elongated tuberous root merging into a short underground stem to form a body to 30 × 15 cm.   Branches numerous and densely clustered, 6–30 cm long, 2–4.5 cm wide, obviously (2)3-angled when young but becoming twisted in a tight spiral, obscurely constricted into segments 1.5–2 cm long; angles with irregular prominent tubercles 5–18 mm apart along the margins. Spine shields usually joined in a continuous horny margin, but separate at the constrictions; spines to 2.5 cm long, reduced to c. 2 mm at the constrictions; prickles vestigial.   Leaves minute, fleshy, deciduous. Cymes solitary, simple, with peduncles 2–4 mm long; bracts oblong, c. 2.5 mm long. Cyathia to 4 × 7 mm with cup-shaped involucres; glands transversely oblong, c. 3.5 mm wide, touching, yellow; lobes rounded, c. 1.5 × 1.5 mm, denticulate. Male flowers: bracteoles c. 3.5 mm long, laciniate; stamens c. 6 mm long, well exserted. Female flower: styles c. 2 mm long, joined to halfway, apices spreading, bifid. Capsule c. 1 cm in diameter, obtusely lobed, sessile. Seeds c. 3 × 2.5 mm, subglobose, buff, smooth.

Mozambique. GI: Chicualacuala Distr., between Chókwe (Guijá) and Mapai, fr. 11.v.1944, *Torre* 6629 (LISC).
Also in South Africa (Northern Province). Mopane woodland; c. 100–200 m.

74. **Euphorbia clavigera** N.E. Br. in F.C. **5**, 2: 362 (1915). —White, Dyer & Sloane, Succ. Euphorb. **2**: 767 (1941). —Court, Succ. Fl. South. Africa: 14 (1981). —S. Carter in Kew Bull. **54**: 960 (1999). Type from Swaziland.
  Euphorbia persistens R.A. Dyer in Fl. Pl. South Africa **18**: t. 713 (1938). —White, Dyer & Sloane, Succ. Euphorb. **2**: 761 (1941). Type: Mozambique, Ressano Garcia to Maputo, st. vii.1936, *van der Merwe* E14 in *Nat. Herb. Pretoria* 23395 (PRE, holotype; K).

Caespitose dwarf spiny succulent perennial, with a large, fleshy, tuberous root merging into a short underground stem, to form a body to 30 × 15 cm. Stem simple, with several short underground branches to c. 3 cm long, rebranching ± at ground level; secondary aerial branches erect, 7.5–20 × 1–3 cm, acutely 3–4(5)-angled, usually constricted irregularly into obovate segments c. 2 cm long; angles sinuate with prominent tubercles 5–20 mm apart along the margins. Spine shields 5–8 mm long, triangular, quite separate; spines to 1.5 cm long but much shorter at the constrictions; prickles minute.   Leaves 1 mm long, deltoid, deciduous. Cymes solitary, simple, with cyathia arranged vertically; peduncles 3–10 mm long; bracts c. 2 × 2 mm, deltoid.   Cyathia with cup-shaped involucres to c. 4 × 7 mm; glands transversely oblong, 3–3.5 mm wide, yellow, touching.   Male flowers: bracteoles laciniate; stamens c. 5 mm long. Female flower: styles 4.5 mm long, joined for 1 mm, apices spreading. Capsule obtusely lobed, 1 cm in diameter, sessile. Seeds not seen.

Mozambique. M: between Moamba and Umbelúzi, fl. 9.xii.1940, *Torre* 2242 (COI; LMU; LISC).

Also in South Africa (KwaZulu-Natal) and Swaziland. Open woodland and coastal forest, in rock crevices and sandy soil; c. 50–120 m.

75. **Euphorbia lividiflora** L.C. Leach in Kirkia **4**: 20, photos in tab. VII (1964). —Drummond in Kirkia **10**: 253 (1975). —Coates-Palgrave, Trees Southern Africa: 449 (1977). —Court, Succ. Fl. South. Africa: 26 & 27 (1981). —Hargreaves, Succ. Spurges Malawi: 16 (1987). — S. Carter in F.T.E.A., Euphorbiaceae, part 2: 497 (1988). TAB. **78**, fig. B. Type: Mozambique, Nhamatanda Distr., SE of Lake Gambué, c. 29 km south of Muda, st. 28.viii.1961, *Leach* 11129A (SRGH, holotype; COI; G; K; LISC; MAL; PRE).

Much branched succulent, spiny tree to 4(10) m high. Trunk woody, to 12.5(25) cm in diameter; main branches spreading, eventually horizontal, 5-angled, eventually terete; secondary branches in distant whorls, 3–4-angled, c. 2.5 cm in diameter; angles prominent, sinuate with tubercles 1–3 cm apart along the margins. Spine shields to 8 × 2.5 mm, obtriangular, usually including the flowering eye, often decurrent and joined in a cartilaginous ridge on older branches; spines to 5 mm long, often becoming obsolete; prickles minute or absent. Leaves to 13 mm in diameter, ± circular, margins strongly revolute, caducous. Cymes solitary, central cyathium often bisexual, forking several times; peduncle 3–6 mm long; cyme branches 5–9 mm long; bracts 4 × 5 mm, ovate, caducous. Cyathia to 4.5 × 9 mm, with broadly campanulate involucres, yellowish to reddish-purple; glands 4.5 mm wide, transversely elliptic, not quite touching, shiny reddish, turning dark purple. Male flowers numerous; bracteoles filiform; stamens c. 5 mm long, anthers almost black. Female flower: perianth 3-lobed, lobes c. 1.25 mm long, irregularly dentate; styles 2 mm long, free almost to the base, spreading, with scarcely bifid capitate apices. Capsule deeply lobed, truncate, to 9 × 25 mm, exserted on an erect pedicel to 8 mm long. Seed subglobose, c. 4.5 mm long, minutely rugulose.

Zimbabwe. E: Chipinge Distr., c. 40 km southeast of Rupise (Rupisi), Chinyamatika road, st. 12.viii.1971, *Percy-Lancaster* 4 (SRGH). S: Chiredzi Distr., Gonarezhou, upper Muwawa R., st. x.1976, *Sherry* in *GHS* 251213 (SRGH). **Malawi**. S: between Chiunguni Hill and Shire R., Liwonde National Park, fr. 20.iii.1977, *Brummitt, Hargreaves, Seyani & Dudley* 14894 (K; MAL; SRGH). **Mozambique**. N: 5 km south of Pemba (Porto Amélia), st., 15.iv.1960, *Gomes e Sousa* s.n. (K; LISC; PRE; SRGH). MS: Machanga Distr., 8 km south of Divinhe, st. 3.ix.1961, *Leach* 11253 (K; PRE; SRGH). GI: Miramar, 24 km northeast of Inhambane, st. 4.xi.1963, *Leach & Bayliss* 11799 (LISC; SRGH).
Also known near the coast in southeastern Tanzania, and probably occurs also in coastal areas of Mozambique in Zambezia Province. Coastal plains and low altitude inland areas, usually in dense thickets on raised ground in floodplain grasslands; also in open dry woodlands and sometimes bordering on mangrove associations; 5–550 m.
Plants sometimes flower and fruit at a height of only 1.5 m.

76. **Euphorbia grandidens** Haw. in Philos. Mag. J. **66**: 33 (1825). —Sim, For. Fl. Port. E. Africa: 105 (1909). —N.E. Brown in F.C. **5**, 2: 372 (1915). —White, Dyer & Sloane, Succ. Euphorb. **2**: 899 (1941). —Coates-Palgrave, Trees Southern Africa: 446 (1977). —Court, Succ. Fl. South. Africa: 17 (1981). Type from South Africa.

Succulent, spiny tree to 10–16 m high. Trunk with several stout, ascending branches, rebranching with ultimate branchlets in clustered whorls at the apices; secondary branches densely rebranching, deeply 2–4-angled, 1.2–2 cm wide; angles sinuate with tubercles 8–30 mm apart along the margins. Spine shields quite separate, 2–7 × 2 mm, obovate; spines to 0.5–6 mm long; prickles vestigial. Leaves 1 × 1 mm, scale-like, deltoid, deciduous. Cymes solitary, simple, subsessile; bracts 1 × 1.5 mm, scale-like. Cyathia 2.5 × 4–5 mm, with cup-shaped involucres; glands transversely oblong, 1.5–3.5 mm wide, touching, yellowish-green; lobes subquadrate, dentate. Male flowers: bracteoles c. 2.5 mm long; stamens to c. 4 mm long. Female flower: perianth 3-lobed, with lobes 1 mm long; styles 1.5 mm long, joined at the base, with spreading, shortly bifid tips. Capsule deeply lobed, c. 8 mm in diameter, exserted on a curved pedicel to 8 mm long. Seeds c. 2.5 mm in diameter, subglobose, brown, smooth.

Mozambique. M: east of Namaacha Falls, fl. & fr. 15.x.1963, *Leach & Bayliss* 11954 (COI; K; LISC; SRGH); Marracuene Distr., Macaneta, st. 10.vii.1980, *Schäfer* 7181 (BR; K; MAL; SRGH; WAG).

Also in South Africa (KwaZulu-Natal and southeast of Eastern Cape). Open dry woodland on rocky slopes, and dry dune forest margins; 5–550 m.

77. **Euphorbia triangularis** Desf. ex A. Berger, Sukk. Euphorb.: 57 (1907). —N.E. Brown in F.C. **5**, 2: 370 (1915). —White, Dyer & Sloane, Succ. Euphorb. **2**: 891 (1941). —R.A. Dyer in Fl. Pl. Africa **43**: t. 1687 (1974). —Coates-Palgrave, Trees Southern Africa: 452 (1977). — Court, Succ. Fl. South. Africa: 17 (1981). Type from South Africa.

Succulent, spiny tree to 9–18 m high. Trunk simple or with several main branches, sparsely rebranching. Ultimate branches clustered in whorls, ascending-spreading, to 1.3 m long, 3.5–10 cm wide, deeply 3–5-angled, constricted into oblong segments 7.5–30 cm long; angles wing-like, 1.5–4.5 cm wide, 3–4 mm thick, shallowly sinuate with tubercles 8–18 mm apart along the margins. Spine shields c. 8 × 3 mm, obovate and separate, or joined in a narrow horny margin along the angles; spines regularly 3–8 mm long, widely diverging; prickles rudimentary. Leaves c. 6.5 × 5.5 mm, deltoid, deciduous. Cymes 1–3 at each flowering eye, simple, with lateral cyathia vertically arranged; peduncles and cyme branches to 3 mm long; bracts scale-like, 1.5 × 2 mm. Cyathia 3 × 4–5 mm, with cup-shaped involucres; glands transversely oblong, 1.5–2 mm wide, touching, yellow; lobes rounded, 1 × 1 mm, denticulate. Male flowers: bracteoles 3.5 mm long; stamens c. 4.5 mm long. Female flower: perianth reduced to a rim; styles c. 3.5 mm long, joined for two-thirds, apices bifid, recurved. Capsule 6–8 mm in diameter, obtusely lobed, exserted on a curved pedicel to 6 mm long. Seeds 2.5 × 2 mm long, ovoid, dark brown, smooth.

Mozambique. M: Inhaca Island, Ponta Ponduini, fr. 18.vii.1957, *Mogg* 27189 (K; SRGH); Namaacha, fl. 25.viii.1967, *Gomes e Sousa & Balsinhas* 4950 (COI).
Also in South Africa (KwaZulu-Natal and southeast of Eastern Cape) and Swaziland. Open dry woodland on rocky slopes and coastal dunes, in sandy soils; 15–600 m.

78. **Euphorbia confinalis** R.A. Dyer in Bothalia **6**: 222 (1951). —Codd, Trees & Shrubs, Kruger Nat. Park: 98 (1951). —Drummond in Kirkia **10**: 253 (1975). —Coates-Palgrave, Trees Southern Africa: 441 (1977). —Court, Succ. Fl. South. Africa:17 & 25 (1981). Type from South Africa.

Succulent spiny tree to 4.5–8(9.5) m tall. Trunk simple or with 1–several main branches, each with a crown of curved ascending branches. Branches 1–1.5 m long, 3–5 angled, constricted into slender ± parallel-sided segments 5–20 × 2.5–5 cm; angles prominent, 5 mm thick, shallowly sinuate, with tubercles 1–2 cm apart along the margins. Spine shields separate when young, later joined to form a horny margin 2–3 mm wide on the angles, blackish; spines 0.5–12 mm long. Leaves 1 × 1 mm, deltoid, deciduous. Cymes 1–3 in a horizontal line 2–5 mm above the spines, simple, with lateral cyathia vertically arranged; peduncles to 2 mm long, cyme branches to 3 mm long; bracts subquadrate, scale-like. Cyathia 4.5 × 6.5 mm with cup-shaped involucres; glands transversely oblong, 2–2.5 mm wide, greenish-yellow; lobes 1.5 × 1.5 mm, rounded, deeply toothed. Male flowers: bracteoles 3 mm long, laciniate; stamens 4–5 mm long. Female flower: perianth triangular, 3 mm in diameter; styles 1.5–3 mm long, joined to halfway, apices bifid. Capsule deeply lobed, 5 × 8 mm, exserted on a recurved pedicel 5–8 mm long. Seeds c. 2.5 × 2.25 mm, ovoid, brown, smooth.

Branches 3–4-angled; spines to 8 mm long, disintegrating · · · · · · · · · · · · · subsp. *confinalis*
Branches 4–5-angled; spines to 12 mm long, persistent · · · · · · · · · · · · · · · subsp. *rhodesiaca*

Subsp. **confinalis** TAB. **79**, fig. A.

Trunk 3–5-angled, simple or with 1–2 trunk-like branches; terminal branches 3–4(5)-angled. Spine shields becoming corky; spines to 8 mm long, wearing away and later obsolete. Cyathial bracts wider than long, c. 2.5 × 3.5 mm, shorter than the involucre.

Zimbabwe. E: Chiredzi Distr., near Rimbi, 19 km south of Rupise (Rupisi), fl. 12.vii.1971, *Percy-Lancaster* 1 (SRGH). **Mozambique**. MS: Mossurize Distr., Búzi R., near Gogói, st.

23.xi.1960, *Leach & Chase* 10532 (PRE; SRGH). M: 16 km north of Moamba, fl. 28.vi.1964, *Leach* 12263 (COI; G; K; LISC; PRE; SRGH).

Also in South Africa in the Lebombo mountain range bordering Mozambique. Numerous on rocky outcrops and ridges in mixed deciduous woodland; 90–500 m.

Subsp. **rhodesiaca** L.C. Leach in J. S. African Bot. **32**: 174, photos on pages 175 & 176 (1966). —Drummond in Kirkia **10**: 253 (1975). —Court, Succ. Fl. South. Africa: 17 (1981). Type: Zimbabwe, Chivi Distr., 1.5 km south of Chivi (Chibi), fl. & fr. 26.viii.1959, *Leach* 9346 (PRE, holotype; SRGH).

Trunk 5–6-angled, usually with several trunk-like branches; terminal branches 4–5-angled. Spinescence more robust, not disintegrating; spines to 12 mm, persistent. Cyathial bracts longer than wide, c. 4.5 × 3.5 mm, ± equalling the involucre in length.

Zimbabwe. W: south Matopos, Kumalo C.L. (Tribal Land), Hendrik's Pass, fl. 31.v.1970, *Leach & Cannell* 14483 (K; LISC; SRGH). S: Bikita Distr., top of Moodie's Pass, fl. 27.viii.1959, *Leach* 9324 (K; SRGH).

Known only in Zimbabwe. Amongst rocks on granite outcrops and ridges; 980–1400 m.

79. **Euphorbia keithii** R.A. Dyer in Bothalia **6**: 223 (1951). —Court, Succ. Fl. South. Africa: 17 (1981). Type from Swaziland.

Succulent spiny shrub or small tree to 2–6 m tall, with a crown of spreading-ascending branches. Branches 1–2 m long, (3)5(6)-angled, constricted into segments up to 25 × 3–4 cm; angles ± winged, with wings 7–15 mm deep and shallowly sinuate, the tubercles 1–1.75 cm apart along the margins. Spine shields joined in a grey, horny margin c. 2–3 mm wide; spines 5–8 mm long, fairly stout. Leaves 5 mm long, ovate, deciduous. Cymes 1–3 in a horizontal line 5 mm above the spines, simple, with lateral cyathia vertically arranged; peduncles and cyme branches 1–2 mm long; bracts subquadrate, scale-like. Cyathia c. 3 × 4–5 mm with cup-shaped involucres; glands transversely oblong, 2–2.5 mm wide, greenish-yellow; lobes 1.25 × 1.25 mm, subquadrate, fimbriate. Male flowers: bracteoles 3 mm long, laciniate; stamens 5 mm long. Female flower: styles 1.5 mm long, joined to halfway, apices bifid. Capsule 4 × 6–7 mm, obtusely lobed, exserted on a recurved pedicel c. 6 mm long. Seeds subglobose, c. 2.25 × 2 mm, brown, smooth.

Mozambique. M: Namaacha Distr., Goba, Fonte de Libombos, fl. 20.vi.1968, *Carvalho* 1013 (LISC).

Also in Swaziland. In Mozambique known so far only from this record in the Pequenos Libombos. On rocky slopes; c. 150 m.

80. **Euphorbia graniticola** L.C. Leach in Kirkia **4**: 18, photos in tab. V, tab VI (1964). —Court, Succ. Fl. South. Africa: 27 (1981). Type: Mozambique, 16 km SSW of Chimoio (Vila Pery), 19°14'S, 33°23'E, *Leach* 5103; cult. Greendale, Harare, fl. & fr. x.1962, (SRGH, holotype; G; K; LISC; LM; PRE).

Succulent spiny shrub or rarely a small tree to 2(2.75) m high, at first branching from the base, eventually forming a stout greyish trunk to c. 1 m tall and c. 12 cm in diameter with a dense crown of crowded, spreading-ascending branches. Branches to 1 m long, 4–6-angled, constricted into segments; segments to 25 cm long, c. 6 cm in diameter near the base and tapering to 4 cm in diameter towards the apex; angles wing-like to 3 cm broad and 3–5 mm thick, shallowly sinuate, with tubercles 5–15 mm apart along the margins. Spine shields joined, forming a conspicuous whitish, horny margin c. 3 mm wide; spines widely diverging, to 8 mm long at base of segments, shorter towards the apex, often much reduced; prickles minute. Leaves to 34 × 4 mm, linear to narrowly ovate, spreading recurved, sessile, caducous. Cymes 3 at each flowering eye, simple, with lateral cyathia vertically arranged; peduncles and cyme branches c. 1 mm long; bracts 1.5–2.5 × 1 mm, broadly ovate, truncate, denticulate. Cyathia with involucres 3 × 3–3.5 mm, narrowly obconic; glands 1.5–2 mm broad, transversely narrowly elliptic, touching, orange-yellow; lobes 0.75 × 0.75 mm, subquadrate. Male flowers numerous: bracteoles 1.75 mm long, filiform; stamens 4 mm long. Female flower: ovary subglobose, perianth rim-like; styles 1.5–2 mm long, united for c. 1 mm, spreading above, stigmas shortly bifid. Capsule 3 × 5.5–6 mm, acutely lobed, truncate at the base, exserted on a recurved pedicel to 9 mm long. Seeds 2 mm in diameter, globose, smooth, pale yellow-brown, mottled.

Mozambique. MS: Manica Distr., Bandula, fr. 19.x.1944, *Mendonça* 2518 (COI; K; LISC; LMU); 35 km southwest of Garuso, fl. & fr. 6.x.1969, *Leach & Cannell* 14378 (K; LISC; SRGH).

Restricted to the area between Chimoio and Manica. Usually in colonies growing in grass clumps on shallow soil patches on granite slopes; c. 700 m.

This is an extremely slow growing species, related to *E. mlanjeana*, and also to the taller *E. decliviticola* which quickly develops a trunk. These three species are discussed by Leach in J. S. African Bot. **39**: 3–18 (1973).

81. **Euphorbia decliviticola** L.C. Leach in J. S. African Bot. **39**: 13, fig. 8 & photos figs. 9–11 (1973). —Court, Succ. Fl. South. Africa: 27 (1981). Type: Mozambique, Ribáuè, c. 2.5 km south of Posto Agrícola, fl. & fr. 20.vii.1962, *Leach & Schelpe* 11427 (SRGH, holotype; BM; BR; K; LISC; PRE).

Euphorbia graniticola sensu Hargreaves, Succ. Spurges Malawi: 29 (1987), non L.C. Leach.

Small succulent spiny tree or shrub to 3 m high; trunk stout, eventually multi-angled, marked with scars of fallen branches. Branches whorled, mostly simple, spreading arcuate-ascending, c. 60 cm long, constricted into segments; segments to c. 15 × 6 cm, 4–6 winged; wings 2–2.5 cm wide, shallowly sinuate with tubercles 5–10 mm apart along the margins. Spine shields forming a continuous whitish horny margin 1.5–3 mm wide; spines horizontally spreading, to 6 mm long; prickles obsolescent. Leaves to 7 mm long, ovate, fleshy, sessile, caducous. Cymes 1–3 at each flowering eye, randomly arranged, lateral cyathia often branching several times; peduncles and cyme branches c. 2 mm long, 2–3 mm thick; bracts 1 × 1.5 mm, broadly ovate. Cyathia with involucres 2.5–3.25 × 4.5–5 mm, obconic; glands fleshy, separate or barely touching, 1.75–2.5 × 1 mm, yellow; lobes 0.5 × 0.5 mm, subquadrate to transversely oblong, fimbriate. Male flowers c. 30: bracteoles filiform, laciniate, c. 2.5 mm long; stamens c. 4.5 mm long. Female flower: perianth rim-like; styles c. 1 mm long, free almost to the base with spreading, shortly bifid apices. Capsule obtusely lobed, 3–3.5 × 5–5.5 mm, exserted on a recurved pedicel 7 mm long. Seeds subglobose, c. 2 mm in diameter, smooth.

Malawi. S: Tundulu Hill, st. 9.x.1980, *Patel & Hargreaves* 752 (MAL). **Mozambique**. N: Ribáuè Distr., c. 18 km east of Namina, st. 23.v.1961, *Leach & Rutherford-Smith* 10976 (K; LISC; PRE; SRGH). Z: Gurné, by R. Licungo, fl. 9.xi.1967, *Torre & Correia* 16043 (LISC).

Known only in central Mozambique and the southern tip of Malawi. Exposed granite slopes and ridges, 600–800 m.

Related to *E. graniticola* from similar habitats in Manica Distr., but that species is primarily an acaulescent shrub, rarely producing a short trunk, and with simple cymes. *E. decliviticola* is also summer-flowering, whereas *E. graniticola* flowers erratically in late winter to early summer.

82. **Euphorbia mlanjeana** L.C. Leach in J. S. African Bot. **39**: 3, fig. 1 & photos figs. 2–5 (1973). —Hargreaves, Succ. Spurges Malawi: 26 (1987). Type: Malawi, Mulanje (Mlanje) Mt., on slopes above the Little Muloza (Malosa) R., st. 15.viii.1971, *Leach* 14805 (SRGH, holotype; K; LISC; PRE).

Euphorbia decliviticola sensu Hargreaves, Succ. Spurges Malawi: 31 (1987), non L.C. Leach.

Succulent spiny shrub to c. 1 m high. Stem stout, to 10 cm in diameter, very short or often to 50 cm high, with a crown of ascending branches. Branches to 60 cm long, c. 3 cm in diameter, 3–5(6)-angled, sometimes slightly constricted into 2–3 segments; angles ± winged with wings c. 1 cm wide, c. 2 mm thick, sinuate, with tubercles 6–12 mm apart along the margins. Spine shields separate at first, becoming joined with age into a horny margin 4–1.5 mm wide; spines 4–5 mm long, widely diverging; prickles rudimentary. Leaves 3–4 × 1 mm, ovate, deciduous. Cymes solitary, simple, subsessile; bracts c. 1.5 × 1.5 mm, scale-like. Cyathia c. 2.5 × 4–4.5 mm, with cup-shaped involucres; glands to 2 × 0.75 mm, transversely oblong, touching, yellow; lobes c. 1.4 × 1.6 mm, obovate, denticulate. Male flowers c. 15: bracteoles filiform, c. 2.5 mm long; stamens c. 4 mm long. Female flower: styles c. 2.25 mm long, joined to halfway, with spreading bifid apices. Capsule 3 × 7 mm, acutely lobed, base truncate, exserted on a recurved pedicel to 8 mm long. Seeds c. 2 × 2.25 mm, subglobose, brown, smooth.

Malawi. S: Mt. Mulanje, Likhubula (Likabula) to Lichenya Path, st. 7.ii.1981, *Chapman &*
Tawakali 5521 (BR; K; MAL).
Known only from southern Malawi. Exposed granite slopes; 1000–1800 m.

83. **Euphorbia memoralis** R.A. Dyer in Fl. Pl. Africa **29**: t. 1129 (1952). —Drummond in Kirkia
    10: 253 (1975). —Court, Succ. Fl. South. Africa: 26 (1981). TAB. **79**, fig. B. Type:
    Zimbabwe, Mvurwi Range (Umvukwe Hills), 1945, *Christian* 379 cult. in *Nat. Herb. Pretoria*
    28439 (PRE, holotype).

Succulent spiny shrubs or tree-like shrubs to 1.5(3) m high, with a short stout ±
cylindric trunk, to c. 10 cm in diameter. Branches crowded, spreading-ascending,
rebranching, dark grey-green, (4)5–7(8)-angled, constricted into segments,
eventually falling with age; segments subconical to ± circular, 5–15 cm long, 3–4 cm
in diameter; angles ± winged, shallowly sinuate, with tubercles 6–10 mm apart along
the margins. Spine shields forming a ± continuous horny margin along the angles,
2–3 mm wide, sometimes interrupted at the flowering eyes; spines to 6 mm long,
spreading, much reduced at the constrictions; prickles minute or absent. Leaves to
20 × 6–7 mm on young growth, oblong-lanceolate, reduced to 1.5–2 mm long on
older growth, triangular, scale-like, caducous. Cymes solitary, 1–2-forked; peduncle
and cyme branches c. 2 mm long; bracts 1.5 × 2 mm. Cyathia with involucres 2.5–3
× 6–7 mm, campanulate; glands c. 3 mm wide, transversely oblong, touching,
greenish-yellow; lobes c. 2 mm wide, fimbriate. Male flowers: bracteoles c. 2.5 mm
long, laciniate; stamens far exserted, pedicels c. 4 mm long, filaments c. 4 mm long.
Female flower: styles 2–4 mm long, free to below the middle, with bifid tips. Capsule
obtusely lobed, c. 4 × 7 mm, exserted on a recurved pedicel 7–8 mm long. Seed c. 2
mm in diameter, subglobose, smooth.

Zimbabwe. N: Zvimba Distr., Mutorashangu (Mtoroshanga) Pass, st. 15.ix.1963, *Leach &*
Müller 11718 (K; SRGH).
Restricted to the northern parts of the Great Dyke. Exposed stony slopes of chrome-rich
soils; 1450 m.

84. **Euphorbia persistentifolia** L.C. Leach in J. S. African Bot. **31**: 251, fig. & photos on pl. XLIX
    (1965). —Fanshawe, Check List Woody Pl. Zambia Showing Distrib.: 23 (1973). —
    Drummond in Kirkia **10**: 253 (1975). —Court, Succ. Fl. South. Africa: 26 (1981). Type:
    Zambia, Mazabuka Distr., Chirundu, Zambezi R. bank, fl. & fr. 20.ix.1959, cult. & flowered
    in Harare, lfs., vi.1964, *Leach* 9383 (SRGH, holotype; BM; G; K; PRE).
    Euphorbia spp. A *& B* of White, Dyer & Sloane, Succ. Euphorb. **2**: 978 & 980 (1941).

Stout succulent spiny shrub to 2 m or small tree to 3.5 m high. Branches whorled,
spreading-ascending, rebranching, acutely 4–5-angled, 2–5 cm in diameter,
constricted into segments to 40 cm long; angles sinuate, with tubercles 1–2 cm apart
along the margins. Spine shields joined in a whitish horny margin 0.5–2 mm wide;
spines 3–7(15) mm long, diverging; prickles minute. Leaves c. 6–10 × 1.6–2.5 cm,
narrowly obovate, subsessile, persistent for several months. Cymes solitary, 1–2-
forked, peduncle and cyme branches c. 3 mm long; bracts c. 3 mm long, ovate.
Cyathia with involucres 3.5–4 × 5–6 mm, funnel-shaped; glands 2–3 mm wide,
transversely elliptic, greenish-yellow; lobes c. 2.25 mm long, broadly elliptic,
fimbriate. Male flowers 20–25(45): bracteoles c. 3 mm long, fimbriate; pedicels and
filaments each c. 2.5 mm long. Female flower: perianth 3-lobed, c. 2 mm in
diameter; styles c. 2.5 mm long, spreading, recurved with bifid apices. Capsule to 4
× 7 mm, obtusely lobed, subtruncate, exserted on a recurved pedicel c. 7 mm long.
Seeds c. 2.5 × 2 mm, subglobose, smooth, pale brown with irregular blotches.

Zambia. S: Gwembe Distr, c. 29 km WNW of Chirundu, x.1959; cultivated and flowered in
Harare, Greendale, fl. & fr. xi.1961, *Leach* 9506 (BM; K; LISC; PRE; SRGH). **Zimbabwe**. N:
Guruve Distr., Zambezi Escarpment, c. 45 km north of Guruve (Sipolilo), st. 17.iv.1960, *Leach*
& Brunton 9867 (K; PRE; SRGH). C: Shurugwi Distr., Bonsor Mine; cult. Pretoria, fl. & fr.
xi.1942, *Carruthers-Smith* in *Nat. Herb. Pretoria* 28440/41/42 (PRE; SRGH). W: Hwange Distr.,
Inyantue Pens, near the Inyantue R., st. 29.iii.1963, *Leach* 11629 (BM; K; PRE; SRGH).
This species occurs mainly along the Zambezi Valley and its Escarpments from below the
Victoria Falls to the Mavuradonha Mts. Escarpment *Brachystegia* woodland, on rocky slopes;
610–1200 m.
A variable species in habit, leaf size and capsule shape.

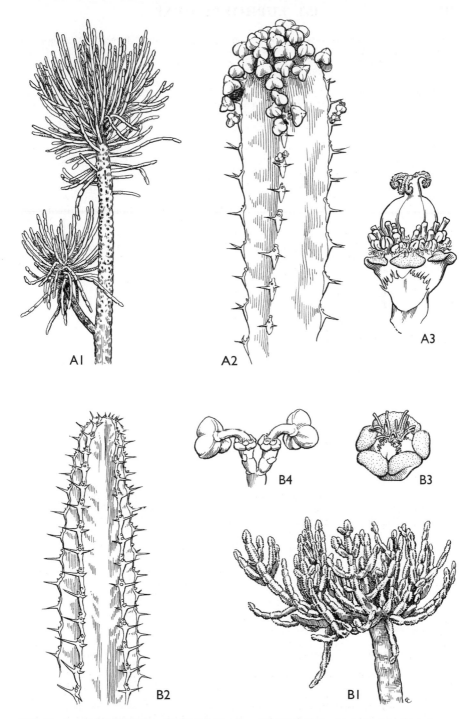

Tab. 79. A. —EUPHORBIA CONFINALIS subsp. CONFINALIS. A1, habit ($\times$ c. $^1/_{60}$), from a photo; A2, apical portion of fruiting branch ($\times$ $^2/_3$); A3, cyathium with developing capsule ($\times$ 4), A2 & A3 from *Leach* 12263. B. —EUPHORBIA MEMORALIS. B1, habit ($\times$ c. $^1/_{30}$), from a photo; B2, apical portion of branch ($\times$ $^2/_3$); B3, cyathium ($\times$ 3); B4, fruiting cyme ($\times$ 1), B2–B4 from *Leach* 11718. Drawn by Eleanor Catherine.

85. **Euphorbia griseola** Pax in Bot. Jahrb. Syst. **34**: 375 (1904). —N.E. Brown in F.T.A. **6**, 1: 578 (1912); in F.C. **5**, 2: 364 (1915). —Burtt Davy, Fl. Pl. Ferns Transvaal: 296 (1932). —White, Dyer & Sloane, Succ. Euphorb. **2**: 774 (1941). —Goodier & Phipps in Kirkia **1**: 58 (1961). —Leach in J. S. African Bot. **33**: 253 (1967). —Court, Succ. Fl. South. Africa: 14, 26 & 27 (1981). —Hargreaves, Succ. Spurges Malawi: 32 (1987); Succ. Botswana: 8 (1990). Type: Botswana, Lobatse, *Marloth* 3413 (B†, holotype); 3 km north of Lobatse, fl. 16.i.1960, *Leach & Noel* 121 (SRGH, neotype selected by Leach 1967; BR; G; K; LISC; PRE).

Spiny succulent shrub to 1 m high, densely branched from the base, or tree-like to 3.5 m high with a central stem. Branches rebranching above, 1–1.5 cm in diameter, 4–6-angled, not segmented to ± distinctly segmented with segments parallel-sided, often variegated between the angles; angles acute, sinuate, with tubercles 8–15 mm apart along the margins. Spine shields joined to form a horny margin along the angles, or spine shields separate; spines 5–7.5 mm long, slender; prickles 0.5–2 mm long. Leaves to 20 × 6 mm, ovate, deciduous. Cymes solitary, simple, subsessile; bracts 1.5 × 1.5 mm, scale-like. Cyathia 3 × 3.5–4 mm, with funnel-shaped involucres; glands 1.5–2 mm wide, transversely oblong, touching, yellow; lobes 1 × 1 mm, rounded, fimbriate. Male flowers: bracteoles 2.5 mm long, laciniate; stamens c. 3.5 mm long. Female flower: styles to 2 mm long, free almost to the base, apices spreading, bifid. Capsule obtusely angled, 5 mm in diameter, exserted on a recurved pedicel to 7 mm long. Seeds 2–2.5 mm in diameter, subglobose.

1. Shrubs tree-like, with a central stem · ii) subsp. *mashonica*
– Shrubs without or with a much reduced central stem · 2
2. Branches occasionally slightly constricted into segments; spine shields usually joined in a continuous horny margin · i) subsp. *griseola*
– Branches distinctly segmented; spine shields usually quite separate · · iii) subsp. *zambiensis*

i) Subsp. **griseola** TAB. **80**, fig. A.

    Euphorbia heterochroma sensu White, Dyer & Sloane, Succ. Euphorb. **2**: 780 (1941) pro parte as to distrib. in Rhodesia and as to fig. 873. —sensu Wild, Dict. Pl. Names: 83 (1952).

    Euphorbia ledienii sensu Suessenguth in Suessenguth & Merxmüller, [Contrib. Fl. Marandellas Distr.] Proc. & Trans. Rhod. Sci. Ass. **43**: 84 (1951).

Shrubs usually less than 1 m high; branches 4–6-angled, occasionally slightly and distantly constricted. Spine shields forming a continuous horny margin along the angles, to 1.5 mm wide, rarely separate.

    Botswana. N: 80 km north of Francistown, fr. 14.i.1960, *Leach & Noel* 27 (K; PRE; SRGH). SE: Mokolodi Hill, fl. 5.vi.1979, *O.J. Hansen* 3228 (K). **Zambia**. S: Livingstone Distr., Victoria Falls, fl. 7.vii.1930, *Hutchinson & Gillett* 3425 (K). **Zimbabwe**. N: Murehwa Distr., 8 km north of Murehwa (Mrewa), fr. 23.vii.1959, *Leach* 9235 (K; PRE; SRGH; ZSS). W: Matobo Distr., Farm Besna Kobila, fl. x.1957, *Miller* 4574 (K; PRE; SRGH). C: Makoni Distr., 8 km southeast of Nyazura (Inyazura), fl. 24.ix.1965, *Leach & Müller* 13138 (BM; BR; K; PRE; SRGH; ZSS). E: Chimanimani Mts., Chikukwa's Kraal, fl. 14.xi.1967, *Mavi* 624 (K; LISC; SRGH). S: near Great Zimbabwe Ruins, fr. 12.i.1961, *Leach* 10711 (BM; BR; G; K; LISC; PRE; SRGH). **Malawi**. S: near Ntaja, fl. 10.viii. 1960, *Leach* 10419 (K; LISC; PRE; SRGH). **Mozambique**. MS: The Corner, st. 20.xii.1957, *Phipps* 841 (SRGH).

    Also in South Africa (Northern Prov.). Among lightly wooded, granite rocks; 850–1700 m.

    Very variable; the form from near Ntaja, in Malawi, has a purplish-red inflorescence, and populations from west of Beitbridge (Zimbabwe) and Francistown (northeast Botswana), are often more than 1 m high.

ii) Subsp. **mashonica** L.C. Leach in J. S. African Bot. **33**: 257, photos on page 259 (1967). — Drummond in Kirkia **10**: 253 (1975). —Hargreaves, Succ. Spurges Malawi: 33 (1987). TAB. **80**, fig. B. Type: Zimbabwe, Mazowe Distr., Tsatse R., 19 km NNW of Concession, fl. & fr. viii.1955, *Leach* 5019 (SRGH, holotype; B; BM; BR; G; K; LISC; LMJ; M; NDO; PRE; ZSS).

Tree-like to 3.5 m high, with a distinct main stem or with several trunk-like branches from the base. Stems often densely branched and rebranched, the main stem 9–12-angled. Spine shields usually joined along the angles.

    Zambia. E: Chadiza Distr., Nsadzu R., fr. 8.x.1958, *Robson* 23 (BM; K; LISC; PRE; SRGH). **Zimbabwe**. N: Shamva Distr., near Mufurudzi (Umfurudzi) and Mazowe (Mazoe) R. confluence, fl. & fr. 7.viii.1959, *Leach* 9267 (BM; K; M; PRE; SRGH; ZSS). C: Goromonzi Distr.,

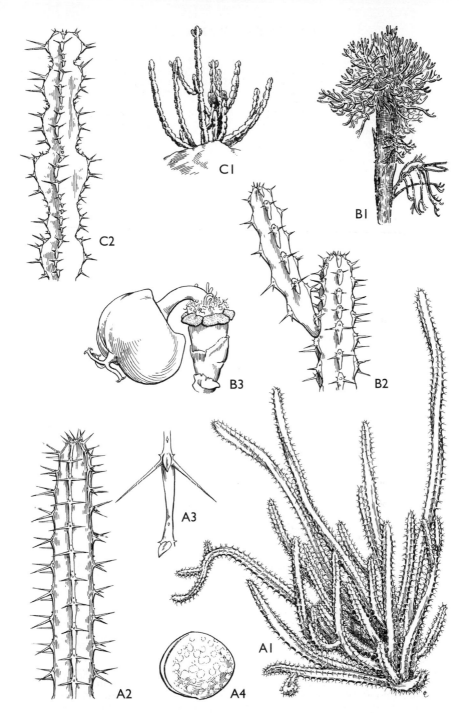

Tab. 80. A. —EUPHORBIA GRISEOLA subsp. GRISEOLA. A1, habit (× c. $^1/_{10}$), from a photo;
A2, apical portion of branch (× $^2/_3$); A3, spinescence (× 2); A4, seed (× 10), A2–A4 from
Leach 9235. B. —EUPHORBIA GRISEOLA subsp. MASHONICA. B1, habit (× c. $^1/_{50}$),
from a photo; B2, apical portion of branch (× $^2/_3$); B3, fruiting cyathium (× 5), B2 & B3
from *Leach* 4019. C. —EUPHORBIA GRISEOLA subsp. ZAMBIENSIS. C1, habit (× c.
$^1/_{25}$), from *Leach & Williamson* 13548; C2, apical portion of branch (× $^2/_3$), from *Leach &*
Brunton 10019. Drawn by Eleanor Catherine.

Ewanrigg, fl. 19.ix.1940, *Christian* 457 (K; PRE). **Malawi**. S: Mpatamanga Gorge, st. 5.v.1960, *Leach & Brunton* 9924 (K; PRE; SRGH). **Mozambique**. T: Changara Distr., Mazoe R., 8 km from Zimbabwe border, fr. 22.ix.1948, *Whellan* s.n. in *GHS* 21931 (K; SRGH).

Known only from the Flora Zambesiaca area. Among granite rocks in *Brachystegia* woodland; 300–1300 m.

iii) Subsp. **zambiensis** L.C. Leach in J. S. African Bot. **33**: 261, photos on page 260 (1967). —
Fanshawe, Check List Woody Pl. Zambia Showing Distrib.: 23 (1973). TAB. **80**, fig. C.
Type: Zambia, c. 56 km east of Kapiri Mposhi, 14.vi.1960, *Leach & Brunton* 10019 (SRGH, holotype; K; LISC; NDO; PRE).

Shrub, ± sparsely branching from the base; branches erect, sparsely rebranched, to 2.5 cm wide, distinctly segmented. Spine shields obovate, separate.

Zambia. N: Chinsali Distr., Shiwa Ngandu, fl. 28.viii.1979, *Chisumpa* 610 (NDO). W: Chingola, fr. 27.x.1957, *Fanshawe* 3821 (K; NDO). C: Kabwe Distr., Kapiri Mposhi, fl. 6.xi.1962, *Fanshawe* 7148 (K; NDO).

Probably also occurs in southeastern Dem. Rep. Congo (Katanga Province). Rocky hillsides, in *Brachystegia* woodland; 1250–1500 m.

86. **Euphorbia rugosiflora** L.C. Leach in Kirkia **13**: 319, figs. 1 & 2 (1990). Type: Zimbabwe,
Chimanimani Mts., The Corner, Muhohwe R., fl. & fr. 26.vi.1988, *D. J. Richards, J. Richards
& Thornton* s.n. in *Leach* 17712 (SRGH, holotype; BM; K; MO; PRE).

Spiny succulent shrub to 40 cm high, branching freely from the base to form dense clumps to 1 m in diameter. Branches ascending, rarely rebranching, (3)4(5)-angled, 10–15 mm in diameter, irregularly constricted; angles sinuate, with tubercles 1–2 cm apart along the margins. Spine shields 5–9 × 3 mm, ovate; spines stout, 4–6 mm long, divergent; prickles minute. Leaves c. 1 mm long, broadly deltoid, recurved, deciduous. Cymes solitary, simple or reduced to a single cyathium, dark red, subsessile. Cyathia with involucres c. 3 × 5 mm; glands transversely oblong, c. 2 mm wide, touching, dark red, conspicuously rugose; lobes c. 1 × 2 mm, broadly fan-shaped, irregularly fimbriate. Male flowers 25–30: bracteoles laciniate, c. 2.25 mm long; stamens to 5 mm long. Female flower: styles c. 2.5 mm long, united to halfway, apices spreading-recurved, shortly bifid. Capsule c. 3.5 × 5 mm, obtusely triangular, red, far exserted on a recurved pedicel 8 mm long. Seeds c. 2 mm in diameter, ± globose, smooth, grey, sometimes brown spotted.

Zimbabwe. E: Chimanimani Mts., The Corner, Muhohwa R., fr. 8.x.1950, *Wild* 3545 (K; LISC; SRGH).

Endemic to the Chimanimani Mts., but possibly occurring also on Mt. Zembe in Mozambique. Sandstone rocks; 1350–1400 m.

87. **Euphorbia richardsiae** L.C. Leach in Kirkia **10**: 392, figs. 1 & 2 (1977). Type: Malawi, near
Mzimba, *Richards* s.n.; cult. Harare, fl. 1.viii.1975, in *Leach* 5239 (SRGH, holotype; BM; BR; K; LISC; PRE).

Spiny succulent densely branching shrublets to c. 15 cm high, or shrubs to 1.25 m high. Branches 4–5-angled, 1–2 cm wide; angles winged, prominently sinuate, with tubercles 4–5 mm high and 8–15 mm apart along the margins. Spine shields to 5–12 × 2.5–3 mm, ovate, decurrent but seldom joined; spines stout, 3–10 mm long, spreading; prickles rudimentary. Leaves c. 2.5–10 mm long, ovate, quickly deciduous. Cymes solitary, simple, subsessile, entirely dark purplish; bracts c. 2.25 × 1.5 mm, ovate. Cyathia c. 2.5 × 4–4.5 mm, with cup-shaped involucres; glands c. 2 mm wide, oblong, closely touching; lobes c. 1.5 × 1.5 mm, fimbriate. Male flowers c. 20: bracteoles filiform, c. 2 mm long; stamens c. 3 mm long. Female flower: styles c. 1.5 mm long, free almost to the base, with widely spreading apices. Capsule 3–3.5 × 4.5 × 5 mm, obtusely lobed, base truncate, exserted on a recurved pedicel c. 10 mm long. Seeds c. 2.25 mm in diameter, broadly ovoid, brown.

Shrublet to 15 cm high; spines to 6 mm long · subsp. *richardsiae*
Shrub to 1.25 m high; spines to 10 mm long · subsp. *robusta*

Subsp. **richardsiae** TAB. **83**, fig. A.

Shrublets to 15 cm high; branches 4–5-angled. Spine shields to 8 × 2.5 mm; spines 3–6 mm long.

Malawi. N: Mzimba Distr., Nkhalapya (Kalapya) Dome, 1968, *Williamson & Simon* 854 (BR; SRGH), cult. Harare, fl. & fr. 13.iv.1971, in *Leach* 14091 (BOL; COI; G; K; MAL; MO; NBG; PRE).
Known only from the type locality. Granite outcrops; 1350–2000 m.

Subsp. **robusta** L.C. Leach in Kirkia **10**: 394 (1977). —Hargreaves, Succ. Spurges Malawi: 37 (1987). Type: Malawi, Mzimba Distr., near Ekwendeni, fl. x.1968, *G. Williamson* 1095 (SRGH, holotype).

Shrubs to 1.25 m high; branches 5-angled. Spinescence robust, with spine shields to 12 × 3 mm; spines to 10 mm long.

Malawi. N: Mzimba Distr., east of Ekwendeni Mission, fl. 1.v.1977, *Pawek* 12754 (K; MAL; SRGH).
Known only from near Ekwendeni. Granite outcrops in *Brachystegia* woodland; 1220–1500 m.

88. **Euphorbia jubata** L.C. Leach in J. S. African Bot. **30**: 7, fig. & photos on pl. IV (1964). — Fanshawe, Check List Woody Pl. Zambia Showing Distrib.: 23 (1973). —Court, Succ. Fl. South. Africa: 29 (1981). Type: Zambia, 3 km southwest of Serenje, fl. & fr. 14.vi.1960, *Leach & Brunton* 10040 (SRGH, holotype; G; K; PRE).

Spiny succulent densely branched and rebranched shrublet c. 15 cm high. Branches 4(5)-angled, c. 1.25 cm wide, deeply furrowed between the angles, shallowly and irregularly constricted into short segments; angles with rounded tubercles 7–12 mm apart along the margins. Spine shields obovate, forming a ± continuous horny margin along the angles, 1–3 mm wide; spines 4–7 mm long, widely diverging; prickles minute, usually ± hooked, sometimes absent. Leaves c. 3.75 × 3.5 mm, broadly ovate, sessile, caducous. Cymes solitary, simple; peduncle 0.75 mm long, 2.5 mm in diameter, with cyme branches stout, c. 2 mm long; bracts c. 2 mm long, subquadrate. Cyathia with cup-shaped involucres, 2–2.5 × 3.5–4 mm; glands transversely oblong, 1.75–2.2 mm wide, touching, yellow to reddish; lobes 1 × 1 mm, broadly elliptic, fimbriate. Male flowers numerous: bracteoles 2.5 mm long, deeply fimbriate; stamens c. 3.75 mm long. Female flower: styles 0.75–1.2 mm long, free almost to the base, with spreading shortly bifid apices. Capsule deeply lobed, c. 3 × 4.5 mm, exserted on a curved reflexed pedicel c. 5 mm long. Seeds subglobose, 1.75–2 mm long, smooth, yellow-brown, often faintly reticulated.

Zambia. N: Mpika Distr., Kaloswe (Koloswe), fl. & fr. 16.vii.1930, *Hutchinson & Gillett* 3771 (K). C: Serenje, fr. 2.xi.1972, *Fanshawe* 11667 (NDO).
Known only from the central watershed ridge in Zambia. Among granite rocks in *Brachystegia–Uapaca* woodland; 1400–1500 m.

89. **Euphorbia knuthii** Pax in Bot. Jahrb. Syst. **34**: 83 (1904). —N.E. Brown in F.C. **5**, 2: 363 (1915). —Pole Evans (ed.) in Fl. Pl. South Africa **9**: pl. 348 (1929). —Burtt Davy, Fl. Pl. Ferns Transvaal: 296 (1932). —White, Dyer & Sloane, Succ. Euphorb. **2**: 732 (1941). — Leach in J. S. African Bot. **39**: 18 (1973). —Court, Succ. Fl. South. Africa: 28 (1981). Type: Mozambique, Ressano Garcia, fl. & fr. 27.xii.1897, *Schlechter* 11949 (K, lectotype; BM; BR; G; PRE; WAG).

Spiny succulent dwarf tuberous-rooted perennial, 1–several-stemmed, often rhizomatous. Branches produced at ground level, simple or sometimes rebranched at the base, seldom above, erect to spreading, 2–4(5)-angled, to 20 cm long or more, 5–9 mm in diameter, distinctly longitudinally whitish-striped between the angles; angles with prominent tubercles 3–5 mm high and 1–3 cm apart along the margins, crowned by the spine shields. Spine shields obovate, decurrent to 1 cm; spines 2–9 mm long, needle-like; prickles minute or absent. Leaves 8–15 × 3.5 mm, lanceolate, caducous. Cymes shortly pedunculate, several times forked with cyme branches to 4 mm long, or cyathia solitary; bracts c. 1.25 × 1 mm, ovate, apiculate. Cyathia with cup-shaped involucres, 2.5 × 3.5–5 mm; glands transversely oblong, 1.5–2.25 mm wide, touching, greenish.   Male flowers: bracteoles c. 2.5 mm long, fimbriate; stamens c. 4 mm long.   Female flower: styles 1.75 mm long, united to halfway, with spreading thickened, shortly bifid apices.   Capsule obtusely lobed, truncate, 4.5 × 6–7 mm, far exserted on a recurved pedicel c. 1 cm long.   Seeds subglobose, c. 3 mm in diameter, smooth, brown.

Plant rhizomatous with several tubers; cymes forking several times · · · · · · · · · subsp. *knuthii*
Plant with a single large tuber; cymes simple · subsp. *johnsonii*

Subsp. **knuthii** —Leach in J. S. African Bot. **39**: 20, photo fig. 12 (1973). TAB. **82**, fig. A.

Habit strongly rhizomatous, with rhizomes producing tubers to 12 × 5 cm, and often with a number of plantlets arising at some distance from the parent plant; branches to 35 cm long, 3–4(5)-angled; cymes often forking several times.

Mozambique. M: 21 km north of Moamba, fl. 29.ix.1963, *Leach & Bayliss* 11749 (K; LISC; PRE; SRGH); 22.5 km north of Catuane, fl. 14.x.1963, *Leach & Bayliss* 11936 (K; LMA; PRE; SRGH).
Also in South Africa (Mpumalanga and KwaZulu-Natal). Distribution is centred around Baía de Maputo extending westwards into Mpumalanga and southwards into KwaZulu-Natal. Floodplain grassland and vleis; 25–200 m.

Subsp. **johnsonii** (N.E. Br.) L.C. Leach in J. S. African Bot. **39**: 20, photo fig. 13 (1973). Type: Mozambique, Machanga Distr., Cherinda, fl. 11.viii.1907, *W.H. Johnson* 271 (K, holotype).
   Euphorbia johnsonii N.E. Br. in F.T.A. **6**, 1: 571 (1911). —White, Dyer & Sloane, Succ. Euphorb. **2**: 736 (1941). —Leach in Kirkia **3**: 34 (1962).

Plants solitary with a single large tuberous root to c. 25 × 8 cm, sometimes ± rhizomatous but without subsidiary tubers or plantlets; branches to 20 cm long, 2–3(4)-angled; cymes usually reduced to solitary cyathia.

Mozambique. MS: Machanga Distr., between R. Búzi and R. Gorongosa (Gorongose), coastal plain, fl. 29.viii.1961, *Leach* 11240 (BM; K; LISC; PRE; SRGH).
Known only from the coastal plain between Rio Búzi and Rio Save, concentrated mainly near Cherinda with a scattered distribution northwards to Ampara. In floodplain grassland often near large saline pans; 4–8 m.

90. **Euphorbia decidua** P.R.O. Bally & L.C. Leach in Kirkia **10**: 293 (1975); in Candollea **18**: 347 (1963) (description without type). —Leach in Bull. Jard. Bot. Belg. **46**: 251 (1976). — Court, Succ. Fl. South. Africa: 27 & 29 (1981). —Hargreaves, Succ. Spurges Malawi: 46 (1987). —S. Carter in F.T.E.A., Euphorbiaceae, part 2: 495 (1988). TAB. **81**, fig. A. Type: Zambia, Mweru Wantipa, fl. 10.i.1944, *Bredo* in *Bally* E271 (K, holotype).

Spiny succulent dwarf perennial, with a much-reduced underground stem merging into an obovoid tuberous root to c. 15–25 × 8 cm, tapering into a long tap-root below. Branches numerous from the stem apex, erect to 15 cm or trailing to 25 cm, simple or sometimes branched at the base, c. 6 mm in diameter, 3–4(6)-angled, withering and finally deciduous in the dry season; angles with prominent tubercles c. 2.5 mm high and 2–10 mm apart along the margins. Spine shields to 3 mm in diameter, rounded, at tips of tubercles, not decurrent; spines 1.5–4.5 mm long, expanded at the base; prickles rudimentary. Leaves 3 × 1 mm, lanceolate, fleshy, deciduous; young plants sometimes producing several leaves from stem apex or branch bases, to c. 5 × 1 cm, lanceolate, tapering into a petiole to 2 cm long. Cymes produced from stem apex after branches have withered, numerous, erect to 3–6 cm long, 1–2-forked, with peduncles 2–4 cm long and cyme branches to 1.5 cm long, cymes sometimes in 2–6-rayed pseudo-umbels on lush specimens with peduncles to 3 cm long; flowering eyes c. 2 mm above the spine shields, rarely producing simple subsessile cymes before the branches wither; bracts c. 2.75 × 2 mm, ovate, denticulate. Cyathia 3 × 3.5 mm, with cup-shaped involucres; glands transversely oblong to 2 mm, touching; lobes 1 × 1 mm, rounded fimbriate. Male flowers: bracteoles 3 mm long, laciniate; stamens 4 mm long. Female flower: styles 1.5 mm long, free to the base with spreading bifid apices. Capsule deeply obtusely

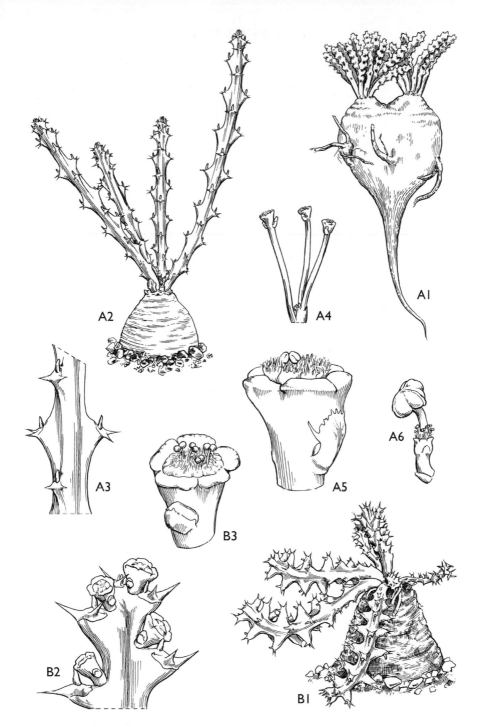

Tab. 81. A. —EUPHORBIA DECIDUA. A1, habit (× ¹/₅); A2, habit with sterile branches (× ²/₃); A3, branch showing spinescence and leaves (× 2), A1–A3 from *Leach & Brunton* 9873; A4, inflorescence (× 1); A5, cyathium (× 6), A4 & A5 from *Whellan* in *GHS* 21938; A6, fruiting cyathium (× 2), from *Bally* E271. B. —EUPHORBIA FANSHAWEI. B1, habit (× ²/₃); B2, apical portion of flowering branch (× 2); B3, cyathium (× 6), B1–B3 from *Radcliffe-Smith, Pope & Goyder.* 5739. Drawn by Eleanor Catherine.

lobed, 3.25 × 5 mm, exserted on a recurved pedicel 6 mm long. Seeds c. 2 mm in diameter, subglobose, smooth.

Zambia. N: Isoka Distr., Mwinimpangala (Mpangara), 9.5 km from Malawi border, northwest of Mafinga Hills, st. 27.xii.1972, *Pawek* 6142 (K; MAL; MO; SRGH). W: Luanshya, fr. 17.ix.1963, *Fanshawe* 7994 (NDO). C: Kabwe, fr. 7.x.1963, *Fanshawe* 8035 (K; NDO). **Zimbabwe**. N: Guruve Distr., 19 km north of Guruve (Sipolilo), st. 16.iv.1960, *Leach & Brunton* 9873 (BR; K; SRGH). **Malawi**. N: Mzimba Distr., 9.5 km northeast of Mzambazi, st. 22.i.1978, *Pawek* 13673 (K; MAL; MO). C: Nkhotakota Distr., near Chipata (Chipala) Hill, st. 16.i.1959, *Robson & Jackson* 1225 (BM; K; SRGH).

Also in southeastern Dem. Rep. Congo and southern Tanzania. *Brachystegia* woodland; 900–1550 m.

91. **Euphorbia fanshawei** L.C. Leach in J. S. African Bot. **39**: 8, fig. 6 & photo figs. 7A & B (1973). —Court, Succ. Fl. South. Africa: 29 (1981). TAB. **81**, fig. B. Type: a specimen cultivated in Harare, fl. 10.iv.1972, *Williamson & Drummond* 1985A, (SRGH, holotype; BR; K), from material collected in Zambia, Kawamba Distr., Ntumbachushi Falls, st. 27.ii.1970, *Williamson & Drummond* 1985.

Spiny succulent dwarf perennial, with a much-reduced underground stem merging into a ± depressed subspherical tuberous root to c. 8 cm in diameter. Branches numerous radiating from stem apex, erect and spreading, simple, to 15 cm long and 4–7 mm in diameter, (4)5–6-angled; angles with very prominent tubercles to 7 mm high and 5–20 mm apart along the margins. Spine shields borne on the upper oblique edge of the tubercles, c. 3 × 3 mm, subquadrate, dark reddish-brown; spines to 4 mm long with expanded bases, widely spreading; prickles obsolete. Leaves to 2.5 mm long, fleshy, subconical, acute, erect, deciduous. Cymes solitary, simple, or occasionally 2-forked or reduced to a single cyathium, peduncles to 5 mm long, cyme branches to 4 mm long; bracts 1 × 1.5 mm, scale-like. Cyathia 2–2.5 × 5 mm, with cup-shaped involucres; glands c. 2 × 1 mm, transversely oblong, spreading, yellow; lobes c. 1 × 2 mm, transversely broadly elliptic, fimbriate. Male flowers c. 20: fascicular bracts finely divided; bracteoles 1.75 mm long, filiform; stamens 3.5 mm long. Female flower: perianth 3-lobed, ovary deeply obtusely lobed, exserted on a recurved pedicel 6 mm long; styles 3–3.5 mm long, shortly united at the base, widely spreading with bifid apices. Capsule and seeds not seen.

Zambia. N: Kawambwa, fl. 16.viii.1973, *Chisumpa* 96 (K; NDO).
Known with certainty only from the type locality, but may also occur in southeastern Dem. Rep. Congo. Plentiful in woodland shade on shallow quartzitic soil, in moist humid conditions; 1100–1280 m.

92. **Euphorbia mwinilungensis** L.C. Leach in Bull. Jard. Bot. Belg. **46**: 246, fig. 2 (1976). Type: Zambia, Mwinilunga Distr., c. 6 km southeast of the Angola border near Mujileshi R., fl. 7.xi.1962, *Richards* 16919 (K, holotype).
Euphorbia decidua sensu Bally & Leach in Candollea **18**: 347 (1963) pro parte as to spec. *Milne-Redhead* 4251 and distribution in Angola.

Spiny succulent dwarf perennial with a large woody tuberous root. Stem subterranean, much reduced, merging imperceptibly with the root, 1–2 cm thick, branching at the apex below ground and producing numerous branches to form large clumps up to 1 m in diameter. Branches crowded, rebranched at the base below ground level, simple, erect, 7–11.5 cm long, 3-angled, c. 5 mm in diameter; angles with prominent tubercles to 2 mm high and 8–25 mm apart along the margins. Spine shields crowning the tubercles, 2–2.5 mm long, ± triangular; spines 3–4 mm long, widely divergent, with expanded bases; prickles minute or absent. Leaves to 2.25 × 1 mm, ovate, sessile, deciduous. Cymes solitary, 1(2)-forked, subsessile; bracts c. 3 × 2.25 mm, broadly ovate. Cyathia 2.5–3 × 6–7 mm, broadly obconic; glands c. 3 mm wide, transversely elliptic, touching, yellow; lobes c. 1.5 × 2 mm, fan-shaped, fimbriate. Male flowers c. 35: bracteoles c. 3 mm long, narrowly cuneate, fimbriate; stamens c. 5.5 mm long. Female flower: styles immature, c. 1.5 mm long, free almost to the base. Capsule and seeds not seen.

Zambia. W: Mwinilunga Distr., top of Kalene Hill, fl. 25.ix.1952, *Angus* 552 (FHO; K).
Also in neighbouring Angola. In deep sand and cracks among granite rocks; 1290–1400 m.

93. **Euphorbia platyrrhiza** L.C. Leach in Bull. Jard. Bot. Belg. **46**: 257, fig. 8 (1976). Type:
Zambia, Kabompo Distr., 45 km west of Kabompo, st. 1.iv.1961, *Drummond & Rutherford-
Smith* 7303 (SRGH, holotype; BR; K).

Spiny succulent dwarf perennial, from a flattish irregularly-shaped woody tuber
to c. 30 cm in diameter and c. 5 cm thick. Stems short, subterranean. Branches
numerous, rebranching near the base, to 22 cm long and 3–5 mm in diameter,
flexuose, 3-angled; angles sinuate-dentate; tubercles 2–4.5 cm apart and 2.5 mm
high along the margins, spirally arranged, sharply or obliquely truncate at the
apex. Spine shields to 2.2 mm, triangular, formed by the expanded bases of the
spines which sometimes remain separated below the leaf scar; spines 2.5–5 mm
long, widely divergent; prickles minute. Leaves 15–25(55) × 5–15 mm, narrowly
obovate, narrowing into a petiole-like base, fleshy, persistent. Cymes solitary,
simple; peduncle and cyme branches 1 mm long; bracts c. 3 × 2.5 mm, oblong.
Cyathia c. 4 × 5 mm, with funnel-shaped involucres; glands 1.5 × 3 mm, ovate,
touching; lobes c. 2 × 2 mm, broadly ovate. Male flowers c. 20: bracteoles c. 2.5 mm
long, filiform; stamens c. 5 mm long. Female flower (immature): pedicel 1.75 mm
long; styles 2–2.5 mm long, free almost to the base, apices shortly bifid. Capsule
and seeds not seen.

Zambia. B: Kabompo Distr., c. 45 km west of Kabompo, st. xii.1969, *Williamson & Simon* in
Leach 14423 (BR; K; SRGH).
Known only from the type locality. Margins of a seasonal pan on Kalahari Sand; c. 800 m.
Measurements of the inflorescence were taken from material of the type cultivated in Harare
by Leach, xi.1969, under his collecting number 14423.

94. **Euphorbia ambroseae** L.C. Leach in Kirkia **4**: 15, photos in tab. I, tab. II (1964). —Court,
Succ. Fl. South. Africa: 27 (1981). —Hargreaves, Succ. Spurges Malawi: 35 (1987). Type:
Mozambique, between Búzi and Gorongosa rivers, fl. & fr. 29.viii.1961, *Leach* 11238
(SRGH, holotype; G; K; LISC; LM; PRE).

A succulent somewhat spiny erect to scrambling shrub up to 2.5 m high; main stem
c. 4 cm in diameter, subcylindric towards the base, angular above, sparingly branched
and rebranched. Branches ascending, occasionally ± constricted into segments,
4(5)-angled, becoming subterete with age, c. 1(2) cm in diameter; angles shallowly
crenate, with tubercles 1–2.5 cm apart along the margins. Spine shields small and
very slender, ± obovate, up to 1 mm wide, narrowly decurrent almost to the flowering
eye below, sometimes united into a horny margin; spines minute or absent; prickles
1–2 mm long, or obsolete. Leaves fleshy, to 10 mm long, orbicular, apiculate,
caducous. Cymes solitary, simple; branches subsessile to c. 2 mm long; bracts c. 2 mm
long, ± elliptic. Cyathia 4 × 5–7 mm, with cup-shaped involucres; glands transversely
3–4 mm oblong-elliptic, yellow-green, outer margin faintly pinkish; lobes 2.5 mm
wide, transversely broadly elliptic, fimbriate. Male flowers: bracteoles c. 3.5 mm long,
broadly laciniate, fimbriate; stamens c. 5 mm long. Female flower: styles 1.75 mm
long, free almost to the base, spreading recurved, apices shortly bifid. Capsule 3–3.5
× 4–5 mm, subglobose, shortly exserted on a pedicel c. 1.5 mm long. Seeds c. 2 mm
long, ellipsoid globose, tuberculate.

Spines and prickles very short · var. *ambroseae*
Spines and prickles obvious · var. *spinosa*

Var. **ambroseae** —Leach in Kirkia **10**: 397 (1977). TAB. **83**, fig. B.

Habit scrambling; spines frequently absent and prickles short.

Mozambique. Z: Mopeia Distr., entre Campo e Mopeia, 30.viii.1949, *Barbosa & Carvalho* 3932
(LMJ). MS: Parque Nacional da Gorongosa (Gorongosa Reserve), Route no. 5, 4.ix.1965,
Macêdo 1254 (LMA). GI: Govuro Distr., c. 16 km south of Nova Mambone, fr. 9.x.1963, *Leach &
Bayliss* 11891 (K; LISC; PRE; SRGH).

Known only from the Mozambique coastal plain from the Zambezi R. southwards to beyond the Save River. In shade, in and on the margins of dense thickets and patches of woodland, on seasonally swampy coastal plain; 7–60 m.

Var. **spinosa** L.C. Leach in Kirkia **10**: 397 (1977). —Hargreaves, Succ. Spurges Malawi: 35 (1987). Type: Mozambique, 3 km north of Save R. and 30 km west of Nova Mambone, *Ambrose* s.n., cult. in Nelspruit, fl. 14.ix.1969, *Leach* 13159 (SRGH, holotype; BR; K; LISC; PRE).

Habit bushy, with erectly spreading branches; spines and prickles all obvious.

Malawi. S: Chikwawa Distr., Lengwe Game Reserve, fl. 20.viii.1976, *Pawek* 11629 (BR; SRGH; WAG); Nsanje Distr., Mwabvi Game Reserve, fr. 4.viii.1975, *Salubeni* 1965 (MAL; SRGH). **Mozambique**. T: Mutarara Distr., opposite Sena, st. vii.1859, *Kirk* s.n. (K); Boruma, near Tete, st. iv.1892, *Menyharth* 1186 (K). MS: Chemba Distr., south bank of Zambezi R., just north of Chemba, 13.vii.1969, *Leach & Cannell* 14329, cultivated in Harare, fl. 8.vii.1970 (K; LISC; PRE; SRGH). GI: Mabote Distr., 45 km from Tessolo on Jofane (Jovane) road, fl. & fr. 11.ix.1973, *Correia & Marques* 3340 (WAG).
More widespread than var. *ambroseae*, sparsely scattered in more open country. Wooded grassland, dry forest and thickets; 40–300 m.

95. **Euphorbia contorta** L.C. Leach in Kirkia **4**: 17, tab III, photos in tab IV (1964). —Court, Succ. Fl. South. Africa: 28 (1981). Type: Mozambique, c. 48 km west of Ribáuè, fl. 16.v.1961, *Leach & Rutherford-Smith* 10889 (SRGH, holotype; G; K; LISC; PRE).

Spiny succulent sparsely branched untidy shrub, c. 1 m high. Branches spreading, sometimes scrambling, usually bent and often twisted, (4)5(7–11)-angled, sometimes constricted into segments to 30 cm long, 2.5–3 cm in diameter; angles acute, with prominent tubercles 8–20 mm apart along the margins. Spine shields 4–9 × 3 mm, obovate, separate; spines 3–10 mm long, narrowly divergent, carmine when young becoming grey; prickles 1–3 mm long. Leaves sessile, fleshy, 8–22 × 7–12 mm, ovate. Cymes solitary, simple, subsessile; branches 1–1.5 mm long; bracts 1.5 × 1.5 mm, subquadrate. Cyathia c. 3.5 × 5 mm, with obconic involucres; glands spreading, transversely oblong, 2.5 mm wide, touching, greenish-yellow, outer margin pinkish; lobes c. 1 × 1.25 mm, transversely elliptic, fimbriate, pale green, speckled red. Male flowers: fascicular bracts broadly flabellate, fimbriate, and bracteoles narrow, c. 3 mm long, laciniate; stamens c. 4.5 mm long. Female flower: styles to 2 mm long, free to the base, spreading recurved, stigmas capitate. Capsule red-brown, c. 3.5 × 4 mm, subglobose, pedicel 1 mm long. Seed ellipsoid, 2.5 × 1.75 mm, verrucose.

Mozambique. N: Cuamba Distr., c. 27 km east of Cuamba (Nova Freixo), fl. 24.v.1961, *Leach & Rutherford-Smith* 10996 (K; SRGH). Z: Gurué Distr., Lioma Mt., 11.vii.1943, *Torre* 5668A (PRE).
Restricted to the granite inselbergs of southern Niassa and Zambezia Provinces. Scattered, in shallow soil on granite slopes; 480–700 m.
Hargreaves has named one of his collections, no. 497 from Malawi, Machinga Distr., Mangombo Hill near Sonje, as *Euphorbia tetracanthoides* (Succ. Spurges Malawi: 42 (1987)), but this identification is extremely unlikely. I have not seen the specimen, but the locality and his sketches suggest it may be *E. contorta.*

96. **Euphorbia baylissii** L.C. Leach in J. S. African Bot. **30**: 213, fig. & photos on pl. XXVII (1964). —Court, Succ. Fl. South. Africa: 27 (1981). Type: Mozambique, Inharrime Distr., Ponta Zavora, fl. & fr. 3.x.1963, *Leach & Bayliss* 11796 (SRGH, holotype; BM; COI; G; K; LISC; MO; PRE).

Spiny succulent shrub-like plant, 0.5–1.8 m high, erect, single stemmed, sparingly branched. Stem occasionally somewhat woody, ± terete, up to 2.25 cm in diameter, stem and branches irregularly constricted with further branches spreading-ascending from the constrictions. Branches 4-angled, to 12–20 mm wide, dark green with central whitish stripe; angles shallowly winged, sinuate with tubercles 4.5 mm high and 8–17 mm apart along the margins. Spine shields to 4 × 1.25 mm, obovate, pale brown; spines spreading, to 4.5 mm long; prickles minute. Leaves c. 1.5 × 1 mm, ovate, caducous. Cymes solitary, simple; peduncle c. 1.5 × 2 mm; cyme branches 2.5–3 mm long; bracts c. 1.25 × 1.25 mm, transversely oblong to subquadrate.

Cyathia 3.5 × 5.5–6 mm, with funnel-shaped involucres, pale yellow tinged purple; glands c. 3 × 1 mm, elliptic, spreading, touching, minutely rugulose, brownish; lobes 2.5 mm wide, transversely elliptic, irregularly denticulate. Male flowers 15: fascicular bracts 2.5 mm long, lacerate, fimbriate; stamens c. 4.5 mm long. Female flower: styles c. 1.25 mm long, free to the base. Capsule obtusely lobed, 3.5–4 × 5.5 mm, shortly exserted on a pedicel 2.5–3 mm long. Seed c. 2 mm long, subglobose, ± densely verrucose.

Mozambique. GI: Inhambane Distr., Miramar, 25 km northeast of Inhambane, fl. & fr. 4.x.1963, *Leach & Bayliss* 11798 (BM; COI; G; K; LISC; MO; PRE; SRGH). M: Marracuene, x.1980, *Schäfer* 7246 (BM; BR; K; SRGH; WAG).
Restricted to coastal areas and off-shore islands. On sand-dunes, often in dune thickets; 0–100 m.

97. **Euphorbia malevola** L.C. Leach in J. S. African Bot. **30**: 1, fig. & photos on pl. I & II (1964). —Fanshawe, Check List Woody Pl. Zambia Showing Distrib.: 23 (1973). —Court, Succ. Fl. South. Africa: 26 (1981). —Hargreaves, Succ. Spurges Malawi: 38 (1987). TAB. **82**, fig. B. Type: Zimbabwe, Mwenezi Distr., 9.5 km south of Runde (Lundi) R. on Harare–Beitbridge road, fl. & fr. 27.viii.1962, *Leach* 5083 (SRGH, holotype; K; LISC; PRE).

Spiny succulent shrub to 1.5 m high, branching densely from the base, sparingly branched above. Branches spreading-ascending, 4(5)-angled, c. 1 cm in diameter, greyish-green with darker mottling along the angles; angles sinuate, with tubercles 8–15 mm apart along the margins. Spine shields 1–2 mm wide, obovate and decurrent to shortly above the flowering eye below, rarely forming a continuous margin, dark purplish-brown; spines needle-like, 5–7 mm long, spreading; prickles c. 2 mm above the spines, 1.5–2 mm long. Leaves c. 1.5 mm long, ovate, caducous. Cymes solitary, simple; peduncle and cyme branches to c. 2 mm long; bracts c. 0.75 × 1 mm, ovate, denticulate. Cyathia 3 × 5–5.5 mm with funnel-shaped involucres; glands 2.25–3 mm wide, transversely oblong, touching, brownish-yellow with a purplish-red outer margin; lobes c. 0.75 × 0.74 mm, denticulate. Male flowers 10: bracteoles c. 2 mm long, fimbriate; stamens 3 mm long. Female flower: styles 3.25–4 mm long, free nearly to the base, apices capitate to emarginate. Capsule to 3.5 × 4 mm, subglobose, dull green marked with purple, sessile. Seeds c. 1.5 mm in diameter, subglobose, verrucose, brownish.

Zambia. S: between Livingstone and Kalomo, st. 10.vii.1930, *Pole Evans* 50, in *Nat. Bot. Gard.* 2793 (PRE; SRGH). **Zimbabwe**. N: 14.5 km west of Mutoko (Mtoko), fl. 23.vii.1959, *Leach* 9239 (COI; PRE; SRGH). W: Hwange (Wankie), near Dutch Reform Church, fl. & fr. 15.viii.1974, *Raymond* 276 (BR; SRGH). C: Mermaid's Pool, fl. 18.vii.1959, *Leach* 9220 (K; LISC; SRGH). E: 9.5 km south of Nyanyadzi, fl. 17.vi.1971, *Percy-Lancaster* 15 (SRGH). S: Hippo Pools, near Runde (Lundi) R. Bridge, fl. xi.1956, *Leach* 5116 (CAH; K; PRE; SRGH). **Malawi**. S: 37 km north of Mangochi, fl. & fr. 10.x.1959, *Leach* 9513 (SRGH). **Mozambique**. T: 38 km from Tete on Chicoa road, st. 27.xii.1965, *Torre & Correia* 13847 (LISC).
Not known elsewhere. Granite rocky outcrops, in open mixed deciduous woodland; 450–1300 m.

98. **Euphorbia dissitispina** L.C. Leach in Kirkia **10**: 394, figs. 3 & 4 (1977). —Court, Succ. Fl. South. Africa: 26 (1981). TAB. **82**, fig. C. Type: Zimbabwe, Bikita Distr., near Umkondo Mine, *Dale* 65; cultivated in Harare, fl. & fr. x.1974, in *Leach* 13240 (SRGH, holotype; BM; BR; K; LISC; PRE).

Spiny succulent much branched shrub to c. 50 cm high. Branches widely spreading, 5–7.5 mm in diameter, 4-angled, becoming subcylindric; angles sinuate to ± straight, with tubercles 10–20 mm apart along the margins. Spine shields 8–10(12) mm long, c. 1.5 mm wide at the apex, narrowing and decurrent below the spines; spines 4–6 mm long, widely spreading, needle-like; prickles c. 2 mm long, widely spreading. Leaves c. 2 × 1 mm, ovate-acuminate, caducous. Cymes solitary, simple; peduncles to 2 mm long; cyme branches to 4 mm long; bracts 1.25–2 × 1–1.25 mm, broadly ovate to oblong. Cyathia c. 3 × 4.5 mm, with funnel-shaped involucres; glands 1 × 2–2.5 mm, transversely oblong, pale greenish-yellow with a narrow reddish outer margin; lobes c. 1.5 × 1.5 mm, obovate, deeply fimbriate. Male flowers 10: bracteoles c. 2.5 mm long, spathulate, laciniate; stamens 5.5 mm long, far exserted,

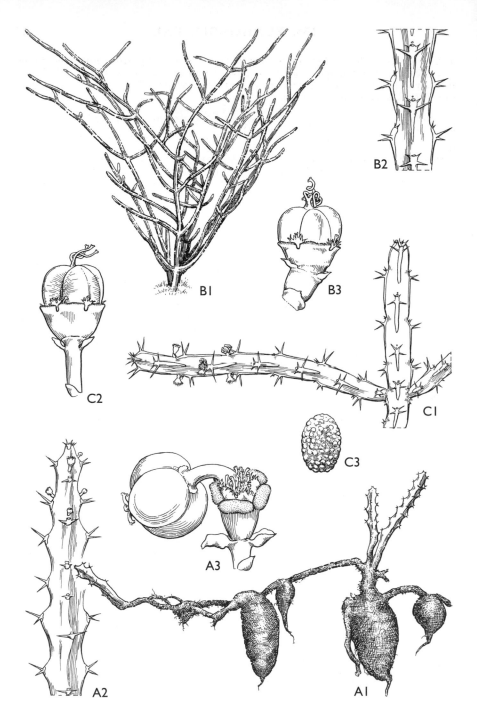

Tab. 82. A. —EUPHORBIA KNUTHII subsp. KNUTHII. A1, habit (× c. $\frac{1}{3}$), from a photo; A2, branch showing spinescence (× 1), from *Leach & Bayliss* 11749; A3, fruiting cyathium (× 4), from *Leach & Bayliss* 11936. B. —EUPHORBIA MALEVOLA. B1, habit (× c. $\frac{1}{16}$), from a photo; B2, portion of branch showing spinescence (× 1); B3, fruiting cyathium (× 4), B2 & B3 from *Leach* 5083. C. —EUPHORBIA DISSITISPINA. C1, branches showing spinescence (× 1); C2, fruiting cyathium (× 4); C3, seed (× 8), C1–C3 from *Dale* 65. Drawn by Eleanor Catherine.

anther locules reddish. Female flower: styles c. 3.5 mm long, spreading, free almost to the base, pale pink, apices minutely bifid. Capsule c. 3.5 × 5 mm, obtusely lobed, scarcely exserted on a fleshy pedicel 1.5 mm long. Seeds c. 2 × 1.5 mm, ± ovoid, pale buff with dense, small, pale warts.

Zimbabwe. S: Bikita Distr., near Umkondo Mine, Save R. Valley, 1966, *Dale* 65; cult. Harare, fl. & fr. ix–xi.1971, in *Leach* 13240 (K; SRGH).
Known only from the type collection, (apparently no longer in cultivation in Harare). Bare stony ground in mixed deciduous woodland; c. 500 m.
Described from a single plant in cultivation, possibly a weak form of *Euphorbia malevola* with more slender branches, shorter spines and other minor differences in measurements. Numerous searches in the region of the type locality have been unsuccessful in finding anything other than *E. malevola*.

99. **Euphorbia cataractarum** S. Carter in Kew Bull. **42**: 379 (1987); in F.T.E.A., Euphorbiaceae, part 2: 514 (1988). Type from Tanzania.

Spiny succulent subshrub to 1.5 m high, sparingly branched. Branches 4-angled, c. 10 mm in diameter, bright green; angles with shallow tubercles to 15 mm apart along the margins. Spine shields c. 2 mm wide, narrowing and decurrent almost to the flowering eye below, pale grey; spines to 6 mm long; prickles to 1.5 mm long. Leaves c. 1.5 × 1.5 mm, deltoid, deciduous. Cymes solitary, simple, subsessile; bracts c. 1.5 × 1.5 mm, subquadrate. Cyathia c. 2.5 × 5 mm, with cup-shaped involucres; glands c. 1 × 2.5 mm, transversely rectangular, touching, yellowish becoming red; lobes c. 1 × 1.5 mm, elliptic. Male flowers: bracteoles c. 2 mm long, denticulate; stamens c. 3 mm long. Female flower: styles 2.5 mm long, joined at the base, apices thickened. Capsule c. 1.8 × 4.5 mm, obtusely lobed, ± sessile. Seeds 2.2 × 1.8 mm, ovoid, minutely verrucose, brown.

Zambia. N: Mbala Distr., Isoko Valley, Mwambezi (Mwambeshi) R., fl. 5.ix.1960, *Richards* 13202 (K).
Also in Tanzania at the southern end of Lake Tanganyika. Amongst rocks in river gorges, characteristically near waterfalls; c. 1200 m.

100. **Euphorbia inundaticola** L.C. Leach in Excelsa No. 15: 18, figs. 12–15 (1992). Type: Zambia, Luangwa Valley, 157 km north of Lundazi, *G. Williamson* 1046; cult. Harare, in *Leach* 14145 (SRGH, holotype; MO; NBG; PRE; UNIN – no type material in K).
Euphorbia schinzii sensu Hargreaves, Succ. Spurges Malawi: 39 (1987) pro parte as to distrib. Mzimba Distr., non Pax.

Spiny succulent subshrub to 1 m high, branching from the base, sparingly branched above. Branches spreading, often trailing with adventitious roots developing, c. 1 cm in diameter, 4-angled; angles sinuate with tubercles 2 mm high and c. 1.5 mm apart along the margins. Spine shields 1.5 mm wide, narrowing and decurrent to 12 mm long, brown to whitish; spines widely diverging, up to 6(8) mm long; prickles c. 2 mm long from 2 mm above the spines. Leaves 1.25 × 1 mm, sessile, caducous. Cymes solitary, simple; peduncles stout, c. 0.5 mm long, cyme branches c. 1.5 mm long; bracts c. 1 × 0.5 mm, broadly triangular. Cyathia c. 3 × 4 mm, with cup-shaped involucres, yellowish to red; glands 2–2.5 × 1 mm, transversely oblong, yellow-reddish; lobes c. 1 × 1 mm, fan-shaped, ± bilobed, irregularly denticulate. Male flowers 5 or 10: fascicular bracts and bracteoles 2.5 mm long, laciniate, fimbriate; stamens 4 mm long, anthers purplish. Female flower: styles c. 3.5 mm long, united for 1.5 mm, stigmas spreading, shortly bifid. Capsule c. 4.5 × 5 mm, obtusely lobed; pedicel c. 1.5 mm long. Seed c. 1 × 1.5 mm, broadly ellipsoid, dark brown, densely verrucose.

Zambia. E: Lundazi Distr., fl. 17.viii.1965, *Fanshawe* 9284 (NDO; SRGH). **Malawi**. N: Mzimba Distr., bridge on S52 between M14 and D172, fl. 26.viii.1980, *Patel* 724 (MAL; SRGH).
Not known elsewhere. In seasonally inundated mopane woodland, on small mounds and amongst rocks; 700–1350 m.

101. **Euphorbia luapulana** L.C. Leach in Excelsa No. 15: 16, fig. 11 (1992). Type: Zambia, Mwense Distr., Musonda Falls, ix.1968, *G. Williamson* 1147 cult. in *Leach* 14147 (SRGH, holotype; MO; NBG; NSW; PRE; UNIN – no type material in K).

Spiny succulent subshrub to 1 m high, branched from the base, sparingly rebranched above. Branches 4-angled, c. 10 mm in diameter, usually tapered towards the base to 5–6 mm in diameter; angles sinuate with tubercles c. 15 mm apart along the margins. Spine shields to 10 mm long, narrowly decurrent ± to the flowering eye below; spines 5–8 mm long, widely divergent to ± deflexed; prickles c. 2 mm above the spines, 1–1.5 mm long, subparallel. Leaves c. 1 × 1.25 mm, deltate, deciduous. Cymes solitary, simple, subsessile; peduncle and cyme branches to 1 mm long, purplish; bracts c. 1 × 1.5 mm, subquadrate, rose-coloured. Cyathia 2.5 × 2.5–4 mm, with funnel-shaped involucres; glands c. 0.75 × 2 mm, elliptic, touching, greenish-yellow; lobes 0.75 × 0.75 mm, ± fan-shaped, shortly bilobed. Male flowers 10–15: bracteoles 2.5 mm long, filiform; stamens c. 3 mm long, anthers pinkish. Female flower: styles c. 2 mm long, shortly united, spreading, recurved, apices thickened. Capsule c. 2.75 × 3.5 mm, very obtusely lobed, eventually bright red; pedicel 1 mm long. Seed 1.75 × 1.5 mm, broadly ellipsoid, verrucose, orange-brown with scattered markings.

Zambia. N: Mwense Distr., Musonda Falls, c. 8 km from the Zambia/Zaire border, 56 km northwest of Mansa (Fort Rosebery), ix.1968, *G. Williamson* 1147; cult. Harare, in *Leach* 14147 (MO; NBG; NSW; PRE; SRGH; UNIN).

Known only from the type collection and described from a plant cultivated in Harare. On rocks close to waterfalls and rapids, in association with *Aloe luapulana* and *Summerhaysia zambesiaca*; c. 1140 m.

No material of this species has been seen by me. The Musonda Falls are on the Luongo R., a tributary of the Luapula R.

102. **Euphorbia speciosa** L.C. Leach in Excelsa No. 15: 19, figs. 16 & 17 (1992). Type: Zambia, Chishimba Falls, *Williamson & Drummond* 2000; cult. Harare in *Leach* 14851 (SRGH, holotype; BR; MO; PRE – no type material in K).
   Euphorbia perplexa var. *kasamana* L.C. Leach in Excelsa No. 15: 16 (1992) pro parte as to spec. *Chisumpa* C99.

Spiny succulent stout shrublet c. 30–90 cm high, sparingly branched from a main stem 2–2.5 cm in diameter at the base. Branches sharply 4(5)-angled, 1–1.5 cm in diameter, blue-green, conspicuously darker striped along the angles; angles sinuate with tubercles c. 4 mm high and 10–12 mm apart along the margins. Spine shields 6–15 mm long, decurrent almost to the flowering eye below; spines 5–8 mm long; prickles to 1 mm long, c. 2 mm above the spines. Leaves 1 × 1 mm, deltoid, deciduous. Cymes solitary, simple, subsessile; cyme branches c. 1 mm long; bracts 1.25–1.75 × 1–1.5 mm, purplish. Cyathia c. 3 × 4.5 mm, with funnel-shaped involucres; glands 0.75–1 × 2–2.5 mm, transversely oblong, touching, orange-red with darker margins; lobes c. 1.3 × 1 mm, broadly cuneate, bright maroon-red. Male flowers 15, far exserted: bracteoles filiform, c. 2.5 mm long; stamens c. 5.25 mm long, anther cells bright maroon. Female flower: styles c. 2 mm long, free almost to the base, bright maroon with yellowish capitate stigmas. Capsule c. 3.5 × 4.5 mm, obtusely lobed, subsessile. Seeds c. 1.9 × 1.4 mm, ellipsoid, densely verrucose.

Zambia. N: Chishimba Falls, fl. 10.ix.1958, *Fanshawe* 4783 (FHO; K; NDO).

Not known elsewhere. In cracks in rock pavement; c. 1450 m.

103. **Euphorbia perplexa** L.C. Leach in Excelsa No. 15: 15, fig. 9 (1992). Type: Zambia, Mbala Distr., Hill above Ndundu, fl. 11.ix.1963, *Richards* 18176 (K, holotype).
   Euphorbia cataractarum sensu S. Carter in Kew Bull. **42**: 380 (1987) pro parte as to *Whellan* s.n. cult. in *Leach* 13148 & 14806.
   Euphorbia nyassae subsp. *mentiens* S. Carter in Kew Bull. **42**: 379 (1987). Type: Zambia, Mbala Distr., Sunzu Mountain, fr. 28.viii.1960, *Richards* 13168 (K, holotype) pro parte excl. *Richards* 13644 & 19280.

Spiny succulent shrublet, erect to 75 cm tall, or with branches semi-prostrate to 1 m long. Branches acutely 4-angled, 10–15 mm in diameter; angles sinuate with prominent tubercles 10–15 mm apart along the margins. Spine shields 6–9 × 3–4 mm, obovate, decurrent to shortly above the flowering eye below; spines sturdy, variable in length, to 5–7 mm long; prickles to 1 mm long and 2 mm above the spines. Leaves 1 × 1 mm, deltoid, deciduous. Cymes solitary, simple, subsessile; bracts c. 1.25–1.75 × 1.25 mm, dentate. Cyathia 2–2.5 × 4 mm, with cup-shaped involucres; glands c. 2–2.5 × 1.25 mm, transversely oblong, touching, yellow; lobes c. 1.25 × 1.25 mm, rounded, denticulate, red. Male flowers 10–15: bracteoles c. 1.5 mm long, filiform; stamens c. 3.5 mm long. Female flower: styles 2.5 mm long, free almost to the base, spreading, apices shortly bifid. Capsule c. 2.5 × 3.75 mm, obtusely lobed, reddish-brown. Seeds c. 1.5 × 1.25 mm, ovoid, lightly verrucose.

Tubercles of the branch angles prominent; spines variable in length, up to 7 mm long · · · · ·
· var. *perplexa*
Tubercles of the branch angles very prominent; spines needle-like, up to 10 mm long · · · · · ·
· var. *kasamana*

Var. **perplexa**

Branch angles prominently sinuate; spine shields to 7.5 × 3.5 mm; spines to 7 mm long, very variable in length; bracts c. 1.25 × 1.25 mm, subquadrate.

Zambia. N: Mbala Distr., Sunzu Mt., fr. 28.viii.1960, *Richards* 13168 (K).
Known only from hills in the Sunzu–Ndundu area. On exposed granite rocks and laterite, also in sandy soil; 1740–2067 m.

Var. **kasamana** L.C. Leach in Excelsa No. 15: 16 (1992). Type: Zambia, north of Kasama, fr. iv.1971, *G. Williamson* s.n.; cult. in Harare in *Leach* 14761 (SRGH, holotype; BR; LISC; MO; NBG; NSW; PRE; UNIN – no type material in K).

Branch angles very prominently sinuate; spine shields to 9 × 4 mm; spines needle-like to 10 mm long; bracts c. 1.75 × 1.25 mm.

Zambia. N: north of Kasama (locality imprecise), fr. iv.1971, *G. Williamson* s.n.; cult. in Harare, in *Leach* 14761 (SRGH, holotype; BR; LISC; MO; NBG; NSW; PRE; UNIN).
Known only from the type, a cultivated plant with no details of its wild locality or habitat. The specimen *Chisumpa* C99 from Kasama township, cited by Leach, has longer spine shields and shorter spines and is identical to others from the same locality and described by him as *Euphorbia speciosa*.

104. **Euphorbia dedzana** L.C. Leach in Excelsa No. 15: 20 (1992). Type: Malawi, Dedza Distr., Kanguli Hill, fl. 18.viii.1971, *Leach & P. Royle* 14826 (SRGH, holotype; K; MO; PRE)*.
  Euphorbia malevola sensu Leach in J. S. African Bot. **30**: 1 (1964), pro parte as to *Greenway* 8814.
  Euphorbia schinzii sensu Hargreaves, Succ. Spurges Malawi: 39 (1987) pro parte as to the Dedza distribution, non Pax.

Spiny succulent shrubs to 30 cm high, densely branched, scrambling and spreading by prostrate ± buried branches to form large patches to 1 m or more across, the branches sometimes establishing separate plants. Branches prostrate to suberect, 30–75 cm long, 8–10 mm in diameter, 4-angled, yellowish to grey with the angles ± purplish; angles with shallow tubercles 12–14 mm apart along the margins. Spine shields c. 8 × 1.5 mm, narrowly triangular, brown; spines 7–9 mm long; prickles c. 2 mm long, 1–1.5 mm above the spines, widely diverging. Leaves 1 × 1 mm, deltoid, deciduous. Cymes solitary, simple; peduncles to 1.5 mm long, cyme branches to 1.25 mm long; bracts c. 1.5 × 2 mm, semicircular. Cyathia 2.5 × 4.5 mm, with cup-shaped involucres; glands 2.25 × 1.25 mm, transversely oblong-elliptic, touching, spreading, yellow; lobes 1.5 × 1.5 mm, denticulate, yellow. Male flowers 7–9, far exserted: bracteoles c. 2 mm long, filiform, fimbriate; stamens 4 mm long. Female flower: styles c. 4 mm long, united for c. 1 mm, then spreading and recurved, yellowish, apices thickened. Capsule 3.25 × 4.5 mm, obtusely lobed, subsessile. Seeds 1.3 × 1.8 mm, broadly ovoid, verrucose, brown with irregular blotches.

---

* Although the type locality is given in the protologue as Dedza Mountain the label data records the specimen as having been collected from "NW of granite hill Kanguli, just north of Angoni Highlands Hotel". As far as can be ascertained there is no locality by the name of Kanguli on the southern slopes of Dedza Mountain, there is however, a Kongoli Hill some 11 km NW of Dedza. It is here considered, therefore, that the type locality is in fact Kongoli Hill within the Chongoni Forest Reserve, and not Dedza Mt.

Malawi. C: Dedza Distr., north of Angoni Highlands Hotel, fl. 30.ix.1954, *Greenway* 8814 (EA; K; SRGH).

Known only from Dedza Mt. and nearby hills. Rocky outcrops, in rock crevices, stunted *Brachystegia–Uapaca–Monotes* woodland; 1600–1800 m.

105. **Euphorbia tholicola** L.C. Leach in Excelsa No. 15: 21 (1992). Type: Malawi, 8 km north of Dedza, fl. 19.viii.1971, *Leach & P. Royle* 14828 (SRGH, holotype; K; MO; PRE).

    Euphorbia schinzii sensu Hargreaves, Succ. Spurges Malawi: 39 (1987), pro parte as to distribution north of Dedza and Dzalanyama Forest Reserve, non Pax.

Spiny succulent shrublet to 15 cm high, densely tufted from the base, sparingly branched above, forming loose cushions. Branches suberect, acutely 4-angled, 6–10 mm in diameter, dull green, darker striped along the angles; angles sinuate with tubercles 2–2.5 mm high and c. 14 mm apart along the margins. Spine shields c. 9 × 1.5 mm, obovate, decurrent to ± halfway to the flowering eye below; spines 4–5 mm long, blackish; prickles 1–2(2.5) mm long, c. 2 mm above the spines. Leaves 1 × 1 mm, deltoid, deciduous. Cymes solitary, simple; peduncle and cyme branches c. 1.5 mm long; bracts c. 2 × 1–1.25 mm, oblong. Cyathia c. 3 × 3.5–4 mm with funnel-shaped involucres; glands 2–2.5 × 1 mm, transversely elliptic, just touching, deep orange; lobes c. 1 × 1.5 mm, transversely elliptic. Male flowers 7–8, far exserted: bracteoles c. 2 mm long, laciniate; stamens c. 5 mm long, anther cells maroon. Female flower: styles c. 3 mm long, united for 1 mm, spreading recurved with bifid tips. Capsule c. 3.5 × 4.5 mm, obtusely lobed, sessile. Seed c. 1.7–2.2 × 1.2–1.7 mm, ellipsoid, brownish, lightly verrucose.

    Zambia. E: Chipata, fl. & fr. 27.vi.1960, *Grout* 229 (FHO). **Malawi**. C: Lilongwe Distr., Dzalanyama Forest Reserve, south of ranch house, st. 26.iii.1977, *Brummitt, Seyani & Patel* 14935 (K; MAL).

    Not known elsewhere. On rocks; 1100–1600 m.

    Leach described this species from a single collection, but specimens from the same area of Dedza Distr., and specimens from Lilongwe Distr., can be identified with it. Together with the type, they appear to be extremely similar to the two collections from a single locality on Dedza Mt., named by him as *E. dedzana*, which differs primarily in the shape of the spine shield. However, the single collection from Zambia is of a smaller, tufted plant, with distinctly longer prickles c. 2.5 mm, and is included here with reservation.

106. **Euphorbia limpopoana** L.C. Leach ex S. Carter in Kew Bull. **54**: 960 (1999). Type: Zimbabwe, Fulton's Drift, 25.5 km NNW of Beitbridge, fl. & fr. 6.ix.1963, *Leach* 11582A (SRGH, holotype).

    Euphorbia malevola subsp. *bechuanica* L.C. Leach in J. S. African Bot. **30**: 6, photo on pl. III (1964). —Hargreaves, Succ. Botswana: 8 (1990). Type: Botswana, halfway between Palapye and Francistown, fl. 1942, *Obermeyer* s.n. (PRE, holotype; K).

Succulent spiny densely tufted shrublet to c. 20–40 cm high; rootstock tuberous fleshy, merging into a short stem below ground with numerous branches arising at or below ground level to form a compact clump to 2 m in diameter. Branches to c. 50 cm long, sharply 4-angled, 1–2 cm in diameter, sparsely rebranching, with a darker stripe along the angles; angles with shallow tubercles c. 1.5–2 cm apart along the margins. Spine shields to 12 × 2 mm, narrowly obovate, decurrent to ± halfway or almost touching the flowering eye below, brown becoming grey; spines stout, 8–17 mm long, widely diverging, prickles 0.5–2 mm long. Leaves 1 × 1 mm, deltoid, deciduous. Cymes solitary, simple, subsessile; bracts c. 1.75 × 1.5 mm, oblong. Cyathia c. 3 × 5 mm, with cup-shaped involucres; glands c. 1 × 2.75 mm, transversely oblong, bright yellow; lobes 1 × 1.5 mm, rounded, denticulate. Male flowers: bracteoles 2.5 mm long, fimbriate; stamens 4 mm long. Female flower: styles 3 mm long, joined at the base, apices minutely bifid. Capsule 3 × 4 mm, obtusely lobed, sessile, purplish. Seeds 2.3 × 1.7 mm, ovoid, verrucose.

    Botswana. SE: Serowe, fr. 1.x.1964, *Leach & Bayliss* 12514 (SRGH). **Zimbabwe**. W: Umguza Distr., Nyamandlhovu, near Umguza R., fl. 1.ix.1955, *Plowes* 1876 (K; SRGH). E: Chipinge Distr., 1 km from Tanganda R. on Mutare (Umtali) to Chipinge (Chipinga) road, fl. 18.viii.1961, *Methuen* 105 (K). S: Chiredzi Distr., Chingedziwa Store, Matibi No. 2 C.L. (T.T.L.), fl. 3.x.1972, *Kelly* 540, in *Leach* 14927 (SRGH). **Mozambique**. GI: Massingir, fl. 6.xii.1980, *Schäfer* 7174 (BR; K; WAG).

Also in South Africa (Northern Prov. and Mpumalanga). On rocky hills and sandy soils in mopane woodland; 140–1200 m.

107. **Euphorbia schinzii** Pax in Bull. Herb. Boissier **6**: 739 (1898); in F.T.A. **6**, 1: 567 (1911); in F.C. **5**, 2: 364 (1915). —Burtt Davy, Fl. Pl. Ferns Transvaal: 296 (1932). —Pole Evans (ed.) in Fl. Pl. South Africa **14**: pl. 523 (1934). —White, Dyer & Sloane, Succ. Euphorb. **2**: 743 (1941). —Court, Succ. Fl. South. Africa: 13 & 26 (1981). —Hargreaves, Succ. Botswana: 8 (1990). TAB. **83**, fig. C. Type from South Africa.

Spiny succulent dwarf shrublet to 33 cm high; rootstock tuberous fleshy, merging into a short stem with numerous rhizomatous branches arising densely at and below ground level, giving rise to further plantlets and forming clumps to 1 m in diameter. Branches to 30 cm long or more, rebranching, (3)4(5)-angled, to 1 cm in diameter; angles with prominent tubercles c. 1 cm apart along the margins. Spine shields c. 7 × 1.5 mm, narrowly obovate, decurrent to ± halfway to the flowering eye below, dark greyish-brown; spines 5–12 mm long, widely diverging; prickles rudimentary or to 1 mm long, c. 1.5 mm above the spines. Leaves 1 × 1.5 mm, deciduous. Cymes solitary, simple, subsessile; bracts c. 1.5 × 1 mm, oblong. Cyathia 2 × 3 mm, with cup-shaped involucres; glands 1–2 mm wide, transversely oblong, bright yellow; lobes 1.5 mm wide, broadly elliptic, denticulate. Male flowers: bracteoles 2 mm long, fimbriate; stamens 3 mm long. Female flower: styles 2–2.5 mm long, joined at the base, apices minutely bifid. Capsule obtusely lobed, c. 3 × 4 mm, sessile. Seeds ovoid, c. 2.25 × 1.5 mm, verrucose.

Botswana. SE: 35 km from Lobatse on Gaborone road, fl. 7.viii.1977, *O.J. Hansen* 3149 (C; GAB; K; PRE; SRGH).
Also in South Africa (North-West Prov., North Prov. and Gauteng). Rocky ground in open woodland; c. 1200 m.
Collections from around, and south of, Bulawayo may represent distinct taxa and require further investigation: *E.B. Best* 631 (K; SRGH) from c. 17 km on the old Bulawayo to Gwanda road, with terete branches and short weak spines; *Miller* 2968 (SRGH) and 6007 (BR; SRGH), from Farm Besna Kobila, south of Bulawayo, with very short branches and a particularly sturdy spinescence.

108. **Euphorbia venteri** L.C. Leach ex R. H. Archer & S. Carter in Fl. Pl. Afr. **57**: 86, pl. 2176 (2001). Type: Botswana, 2 km north of Tsamaya near Tshesebe (Tsessebe), c. 45 km north of Francistown, 12.xii.1991, *Venter, Hahn & Archer* 174 (PRE, holotype; K; UNIN).

Spiny succulent dwarf shrublet, with a short stem merging into a fleshy tuberous root, with numerous branches forming a compact clump. Branches 10–15 cm long, up to c. 1 cm in diameter, ± terete, greyish-green, often pale-striped; tubercles in 4 series, prominent, c. 1 cm apart. Spine shields c. 6.5 × 3 mm, obovate, decurrent to ± halfway to the flowering eye below, greyish-brown; spines 3–6 mm long, widely diverging; prickles rudimentary or up to 0.5 mm long, c. 1 mm above the spines. Leaves 1 × 1 mm, deltoid, deciduous. Cymes solitary, simple or sometimes 2-forked, subsessile; bracts c. 1.25 × 1.25 mm, subquadrate. Cyathia 3 × 3 mm, with funnel-shaped involucres; glands c. 1.5 mm wide, transversely elliptic, green becoming brownish; lobes c. 1 mm wide, subquadrate, denticulate. Male flowers: bracteoles c. 1.25 mm long, fimbriate; stamens 3 mm long. Female flower: styles 2 mm long, joined at the base, spreading at the apex, bifid. Capsule acutely lobed, c. 3 × 4 mm, subsessile. Seeds not seen.

Botswana. N: 2 km north of Tsamaya near Tshesebe (Tsessebe), fl. 12.xii.1991, *Venter, Hahn & Archer* 174 (PRE).
Reported also from Foley Siding by Hargreaves (pers. com.), 2.vi.1993, *Hargreaves* 6700 (GAB; PRE); clay soil in mopane woodland, c. 1220 m.

109. **Euphorbia tortistyla** N.E. Br. in F.T.A. **6**, 1: 569 (1911). Type: Zimbabwe, Masvingo (Victoria), fr. 1909, *Monro* 490 (BM, holotype; K, drawing; SRGH).

Spiny succulent dwarf shrublet to 15 cm high, with densely tufted, stoloniferous branches, forming compact clumps to 50 cm in diameter. Branches 5–8 mm in diameter, very obtusely 4-angled; angles shallowly sinuate with tubercles 6–10 mm

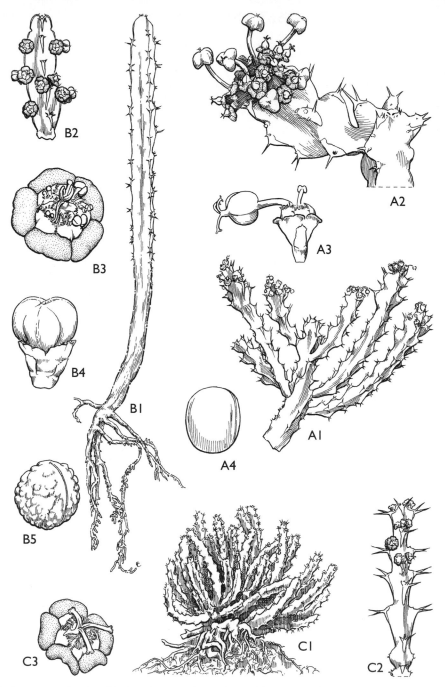

Tab. 83. A. —EUPHORBIA RICHARDSIAE subsp. RICHARDSIAE. A1, fertile branches (×
²/₃), from *Williamson* 503; A2, fruiting branch (× 1), from *Williamson & Simon* 854; A3,
fruiting cyathium (× 4), from *Williamson* 503; A4, seed (× 10), from *Williamson & Simon*
14091. B. —EUPHORBIA AMBROSEAE var. AMBROSEAE. B1, seedling (× ²/₃); B2,
flowering branch (× ²/₃); B3, cyathium (× 4), B1–B3 from *Leach* 11238; B4, fruiting
cyathium (× 4), from *Leach & Bayliss* 11891; B5, seed (× 10), from *Leach* 11238. C. —
EUPHORBIA SCHINZII. C1, habit (× c. ¹/₄), from a photo; C2, flowering branch (× ²/₃);
C3, cyathium (× 4), C2 & C3 from *Leach & Noel* 218. Drawn by Eleanor Catherine.

apart along the margins. Spine shields 3–8 × 2–2.5 mm, obovate, decurrent to less than halfway to the flowering eye below; spines 2–8 mm long; prickles minute to 0.5 mm long, 1–1.5 mm above the spines. Leaves scale-like, 0.5 × 0.75 mm, deltoid, deciduous. Cymes solitary, simple; cyme branches 1–2 mm long; bracts 1.5 × 1 mm, oblong, denticulate. Cyathia c. 2.75 × 3.5 mm, with cup-shaped involucres; glands to 2 mm wide, oblong, touching; lobes subquadrate, 1.25 × 1.5 mm, fimbriate. Male flowers: bracteoles 2.75 mm long; stamens c. 3.5 mm long. Female flower: styles 2.5 mm long, united at the base, spreading, with shortly bifid apices, spirally twisted when dry. Capsule 3.25 × 3.5 mm, obtusely lobed. Seeds 2 × 1.4 mm, ovoid, verrucose.

Zimbabwe. S: 22.5 km south of Gutu on Masvingo (Fort Victoria) road, st. 4.xii.1960, *Leach* 10545 (K; SRGH).
Known only from the Zimbabwe central plateau southern margins. Exposed granite slabs and domes, in rock crevices and outcrop margins in decomposed granite; 1100–1400 m.

110. **Euphorbia acervata** S. Carter in Kew Bull. **54**: 962 (1999). TAB. **84**, fig. A. Type: Zimbabwe, Mvurwi Range (Umvukwe Hills), Mutorashangu (Mtoroshanga) Pass, fl. ix.1940, *Christian* 383 (K, holotype; SRGH).

Spiny succulent dwarf shrublet, with several short stems from a tuberous root, and densely tufted branches forming cushions to 30 cm high and 50 cm in diameter. Branches seldom rebranched, 10 mm in diameter, very obtusely 4-angled; angles with prominent tubercles 3–4 mm high and 10–17 mm apart along the margins. Spine shields 4–8 × 2–3 mm, obtriangular; spines to 8 mm long, divergent; prickles 1–2 mm long, 1.5–2 mm above the spines. Leaves 0.5 × 0.75 mm, deltoid, fleshy, deciduous. Cymes solitary, simple, subsessile; bracts 1.5 × 1.5 mm, subquadrate, denticulate. Cyathia 2.5 × 5 mm, with broadly funnel-shaped involucres; glands 2.25 × 1 mm, oblong, touching; lobes 1.75 × 1 mm, elliptic, denticulate. Male flowers: bracteoles 1.75 mm long, fan-shaped, fimbriate; stamens 3 mm long. Female flower: styles 2 mm long, joined at the base, spirally twisted when dry. Capsule 3.75 × 4.25 mm, sessile. Seeds c. 2 × 1.5 mm, ovoid (only immature seen).

Zimbabwe. N: Zvimba Distr., near Mpinga, 16.ix.1962, *Leach* 11521 (SRGH). C: Kadoma Distr., 1 km north of Ngezi Dam, fl. 10.viii.1988, *Carter & Coates Palgrave* 2686 (K).
Known only from the Great Dyke. In grass among rocks in open woodland; 1240–1450 m.

111. **Euphorbia distinctissima** L.C. Leach in Excelsa No. 15: 12, figs. 4 & 5 (1992). Type: Zambia, c. 76 km south of Mbala, ii.1970, *Williamson & Drummond* 1994; cult. in Harare in *Leach* (SRGH, holotype; BR; MO; PRE; UNIN).

Spiny succulent shrublets to c. 25 cm high, branching freely from the base, forming large clumps. Branches usually simple, up to 23 cm long and 10 mm in diameter, subcylindric; tubercles in 3–5 spiral series, 5–10 mm apart. Spine shields 2.5–7.5 × 1.5–2.5 mm, obtriangular, brown, with a narrow dark margin when young, becoming grey; spines 1–4 mm long, widely diverging, deflexed; prickles 1–2.5 mm long. Leaves 0.75 × 0.75 mm, ovate-acute, deciduous. Cymes solitary, simple, with peduncles 1–2.5 mm long, purplish; cyme branches to 3 mm long; bracts c. 1 × 0.75 mm, ovate to oblong, truncate, maroon. Cyathia 2.5 × 3–3.5 mm, with narrowly funnel-shaped involucres; glands c. 1.3 × 0.5 mm, transversely elliptic, spreading, touching, orange-yellow; lobes c. 1 × 1 mm, ± bilobed, bright cerise. Male flowers c. 10: bracteoles few, c. 2.5 mm long, filiform to very narrowly oblong, denticulate; stamens c. 5 mm long, anther cells dull cerise. Female flower: styles 2.5 mm long, spreading, recurved, ± free to the base, apices shortly bilobed. Capsule 3 × 3.5 mm, obtusely lobed, barely exserted. Seeds c. 1.8 × 1.3 mm, ± ellipsoid, prominently verrucose, brown.

Zambia. N: Mbala Distr., c. 76 km south of Mbala (Abercorn) on road to Mporokoso, fl. 5.ix.1950, *Bullock* 3312 (K; SRGH).
Known only from the type locality (the Senga Hill area in northern Zambia), but according to Leach (in Excelsa No. 15: 12 (1992)) plants to the north of Mpika may belong here. Rocky outcrops in *Brachystegia* woodland; c. 1600–1750 m.

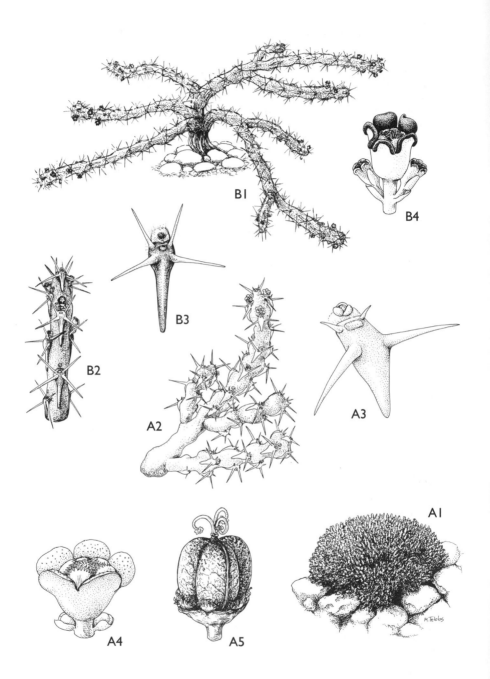

Tab. 84. A. —EUPHORBIA ACERVATA. A1, habit (× c. ¹/₄); A2, branch (× ²/₃); A3, spine shield, with spines, prickles, leaf scar and flowering eye (× 4); A4, immature cyme (× 8), A1–A4 from *Carter & Coates-Palgrave* 2249; A5, capsule (× 8), from *Leach* 11521. B. — EUPHORBIA PLENISPINA. B1, habit (× ²/₃); B2, branch (× 2); B3, spine shield, with spines, prickles, leaf scar and flowering eye (× 4); B4, immature cyme (× 8), B1–B4 from *Carvalho* 1019. Drawn by Margaret Tebbs. From Kew Bull.

112. **Euphorbia isacantha** Pax in Bot. Jahrb. Syst. **34**: 82 (1904). —N.E. Brown in F.T.A. **6**, 1: 575 (1911). —Brenan, Check-list For. Trees Shrubs Tang. Terr.: 212 (1949). —Hargreaves, Succ. Spurges Malawi: 44 (1987). —S. Carter in F.T.E.A., Euphorbiaceae, part 2: 514 (1988). —Leach in Excelsa No. 15: 14 (1992). Type from Tanzania.

Spiny succulent shrublet to c. 25 cm high, intricately branched from the base to form tangled heaps up to 1 m in diameter. Branches to 30 cm long, 8–10 mm in diameter, obtusely 4-angled, pale green; angles shallowly sinuate with tubercles c. 10 mm apart along the margins. Spine shields 5–9 × 1.25–2 mm, decurrent and narrowing slightly but not reaching the flowering eye below, whitish-buff; spines slender, to 4 mm, occasionally obsolescent, widely diverging, usually deflexed; prickles usually slightly longer than the spines, to 5 mm long, c. 2 mm above the spines, divergent to form an X with the spines. Leaves c. 2 mm long, ovate-acute, caducous. Cymes solitary, simple; peduncles to 2.5 mm long, cyme branches to 3 mm long; bracts to 2.25 mm long, ovate, subacute. Cyathia c. 3.5 × 5–6 mm with funnel-shaped involucres; glands c. 1.2 × 2.7 mm, transversely oblong, touching, dull orange, outer margin reddish; lobes c. 1.5 × 1.5 mm, fimbriate, bright magenta. Male flowers 10: bracteoles c. 3 mm long, filiform; stamens c. 5 mm long. Female flower: styles c. 3.5 mm long, free to the base, stigmas spreading, rugulose. Capsule and seeds not seen.

Malawi. N: Karonga Distr., 8 km west of Karonga, Chaminade School, st. 15.iv.1976, *Pawek* 11044 (K; SRGH).
Also in southeastern Tanzania. In wooded grassland and open *Brachystegia* woodland; 550 m.

113. **Euphorbia torta** Pax & K. Hoffm. in Bot. Jahrb. Syst. **45**: 240 (1910). —N.E. Brown in F.T.A. **6**, 1: 568 (1911). —S. Carter in F.T.E.A., Euphorbiaceae, part 2: 511 (1988). Type from Tanzania.

Spiny succulent shrublet, branching densely from the base to form clumps up to c. 15 cm high; root thick, woody, rhizomatous. Branches up to 25 cm long, 4–5 mm in diameter, 4-angled; angles with prominent tubercles to 1.5 cm apart along the margins. Spine shields to 12 mm long, c. 2 × 2 mm at the apex, very slender below the spines, reddish-brown; spines to 3 mm long; prickles to 2.5 mm long. Leaves c. 1.5 × 1 mm, deltoid, deciduous. Cymes solitary; peduncles and cyme branches c. 3 mm long; bracts c. 1.5 × 1.25 mm, ovate. Cyathia c. 3 × 4.5 mm, with cup-shaped involucres; glands c. 1 × 2 mm, transversely oblong, not quite touching; lobes 1 × 1 mm, rounded, denticulate. Male flowers few: bracteoles c. 2.5 mm long, deeply dentate; stamens 4.5 mm long. Female flower: styles 3.5 mm long, joined to nearly halfway, apices slightly thickened. Capsule and seeds not seen.

Zambia. N: Mbala Distr., Mpulungu Escarpment, st. v.1937, *Burtt* 6193 (BM; K).
Also in southeastern Tanzania. *Brachystegia* woodland; 800–1600 m.

114. **Euphorbia plenispina** S. Carter in Kew Bull. **54**: 964 (1999). TAB. **84**, fig. B. Type: Mozambique, between Machaze and Madindire, fl. 28.vii.1968, *Carvalho* 1019 (LISC, holotype).

Spiny succulent shrublet, procumbent to 15 cm high or suberect to c. 50 cm tall, branching from a thickened root. Branches sparsely rebranched, to 8–25 cm long, c. 5 mm in diameter, obtusely 4-angled; angles shallowly sinuate with tubercles 5–10 mm apart along the margins. Spine shields 3–6 × 1.5–2 mm, obovate, decurrent from halfway to almost touching the flowering eye below, red-brown; spines and prickles equal in length, 3–4 mm long, very slender, blackish, divergent to form an X, with prickles 1–1.5 mm above the spines. Leaves 0.75 × 0.75 mm, deltoid, deciduous. Cymes solitary, simple, red; cyme branches 1 mm long; bracts 1 × 0.75 mm, oblong. Cyathia c. 2 × 3 mm, with cup-shaped involucres; glands 1.25 × 0.75 mm, elliptic, separate; lobes 0.75 × 0.5 mm, rounded, denticulate. Male flowers: bracteoles 1.75 mm long, laciniate; stamens immature. Female flowers: capsules and seeds not seen.

Mozambique. MS: Machaze Distr., Save (Sabi) R., 8 km north of Maringua's Village, fl. 23.vi.1950, *Chase* 2241 (BM).

Known only from the area of the type locality (just north of the Save R.). On exposed rocks, among lichens; c. 75–250 m.

115. **Euphorbia whellanii** L.C. Leach in J. S. African Bot. **33**: 247, fig. & photos on page 249 (1967). —Court, Succ. Fl. South. Africa: 29 (1981). —Leach in Excelsa No. 15: 10, fig. 3 (1992). Type: Zambia, Mbala Distr., near Kawimbe Mission, fl. & fr. (no date), *Whellan* 2122 (SRGH, holotype; K; PRE).

    Euphorbia nyassae subsp. *mentiens* S. Carter in Kew Bull. **42**: 379 (1987) pro parte as to *Richards* 13644 (K).

Spiny succulent dwarf perennial, densely branched from a thickened root. Branches usually simple, erect to 17.5 cm tall or trailing to 25 cm long, 8 mm in diameter, obtusely (4)5–7(8)-angled; angles sinuate with tubercles 3–5 mm apart along the margins. Spine shields 2–3 × 1.5–2 mm, obovate; spines 2.5–6 mm long; prickles 1.5–3 mm long, slender, widely diverging to form an X with the spines. Leaves 0.5 mm long, deltoid, soon deciduous. Cymes solitary, simple; peduncles and cyme branches c. 2 mm long; bracts subquadrate, c. 1.25 mm long, dentate. Cyathia c. 2 × 3.5 mm with funnel-shaped involucres; glands transversely oblong to 0.6 × 2 mm, spreading, yellow, separate; lobes rounded, c. 1 × 1 mm, denticulate. Male flowers 10–12: bracteoles 2 mm long, laciniate; stamens 2.5 mm long. Female flower: styles 2 mm long, free almost to the base, spreading, apices capitate. Capsule obtusely lobed, c. 2.5 × 4 mm, sessile. Seeds c. 1.75 × 1.5 mm, subglobose, densely verrucose, whitish, speckled.

Zambia. N: Mbala Distr., near Tanzania border, fr. 12.vi.1957, *Richards* 10082 (K).
Also in adjacent southwestern Tanzania. On exposed granite slabs, in crevices amongst short grasses; c. 1700 m.

116. **Euphorbia debilispina** L.C. Leach in Excelsa No. 15: 13, figs. 6 & 7 (1992). Type: Zambia, 8 km west of Lusaka, *Williamson & Drummond* 1995 (SRGH, holotype; K; MO; PRE).

Spiny succulent dwarf shrublet, forming large densely branched clumps up to c. 12.5 cm high. Branches subcylindric or obtusely subquadrangular, 8–10 mm in diameter, tapering below into a stalk-like base; angles shallowly crenulate with rounded tubercles c. 7 mm apart along the margins. Spine shields 2–4 × 2–3 mm, obtriangular, brownish becoming whitish; spines 2–3 mm long or sometimes rudimentary, widely divergent; prickles 1–2 mm long, weak. Leaves c. 1.25 mm long, ovate-attenuate, fleshy, caducous. Cymes solitary, simple; peduncles to 2 mm long; cyme branches to 3 mm long; bracts c. 1.5 mm long, semi-circular to oblong-obtuse. Cyathia to 2 × 4–4.5 mm, with funnel-shaped involucres; glands spreading, transversely elliptic, c. 2 mm wide, ± touching, yellow; lobes 1 × 1 mm, cuneate, irregularly fimbriate. Male flowers: 10–15, far exserted: fascicular bracts broadly 2-lobed, laciniate, fimbriate, 2.5 mm long; stamens 4.5 mm long. Female flower: styles 2.25–3 mm long, united to c. halfway, spreading above, with capitate or shortly bilobed apices. Capsule c. 3.5 × 4.25 mm, obtusely 3-lobed, scarcely exserted. Seeds c. 2 × 1.25–1.5 mm, ovoid, densely verrucose, brown with white, brown and blackish markings.

Zambia. C: 14 km west of Lusaka, fl. 10.ix.1993, *Merello, Harder & Bland* 920 (K; MO).
Known only from west of Lusaka. On limestone pavement in sparse open deciduous woodland; c. 1250 m.

117. **Euphorbia ramulosa** L.C. Leach in J. S. African Bot. **32**: 176, fig. (1966). —Court, Succ. Fl. South. Africa: 28 (1981). Type: Mozambique, 60 km west of Nampula, st. vii.1962, *Leach & Schelpe* 11440 (SRGH, holotype; K; LISC; PRE).

Spiny succulent shrublet to 15–40 cm high, densely branched. Branches rigid, up to 17 cm long, 4-angled, 10–20 mm in diameter; angles with prominent tubercles to 2.5 mm high, 8–12 mm apart along the margins. Spine shields obovate, forming a continuous horny margin along the angles c. 1.5 mm wide; spines stout, 5–8 mm long, spreading; prickles minute or absent, c. 2 mm above the spines. Leaves 1 × 0.75 mm, caducous. Cymes solitary, simple; peduncles to 3 mm long, cyme branches 1.75–4.5 mm long; bracts 1.5 mm long, ovate, denticulate. Cyathia with involucres 1.75–2.75 × 3.5–5.25 mm, broadly funnel-shaped; glands 1.5–2.75 mm wide, transversely oblong, touching, greenish-yellow; lobes 1.5 × 2 mm, broadly cuneate, fimbriate. Male flowers 25: bracteoles c. 2.5 mm long, filiform-fimbriate; stamens c. 4 mm long, reddish, with purplish anther cells. Female flower: styles 1–1.2 mm long, free nearly to the base, purplish, with yellowish capitate apices. Capsule c. 3.25 × 3 mm, obtusely lobed, angles and sutures purplish; pedicel c. 1.5 mm long. Seeds c. 1.75 × 1.2 mm, ellipsoid, conspicuously verrucose, greyish-brown.

Mozambique. N: 79 km from Marrupa towards Mecula, st. 11.viii.1981, *Jansen, de Koning & de Wilde* 193 (BR; K). Z: Ile Mts., c. 3 km from Ile (Errego), st. 3.iii.1966, *Torre & Correia* 14998 (LISC).
Not known elsewhere. Exposed granite slabs and domes, in crevices, in open woodland; 395–950 m.

118. **Euphorbia corniculata** R.A. Dyer in Fl. Pl. Africa **27**: t. 1076 (1949). —Court, Succ. Fl. South. Africa: 28 (1981). Type: Mozambique, near Nampula, fl. i.1944, *Gomes e Sousa* 3339 cult. in *NH* 27271 (PRE, holotype).

Spiny succulent shrublet to 15 cm high, densely branching from the base to form clumps up to 80 cm in diameter. Branches spreading, 10–15 mm in diameter, terete, with 6–8 angles or ridges separated by grooves; angles with prominent tubercles 2 mm high and up to 8–12 mm apart along the margins. Spine shields joined to form continuous, sinuous, horny, metallic-grey ridges 5 mm wide, separated by grooves; spines to 8 mm long, variable in length; prickles obsolete. Leaves 0.75 × 0.75 mm, deciduous. Cymes solitary, simple; peduncles and cyme branches 1–2.5 mm long; bracts 1.5 mm long, oblong, denticulate. Cyathia 2 × 3.5 mm, with cup-shaped involucres; glands 1.5 mm wide, transversely oblong, touching, dark red; lobes 1 × 1 mm, subquadrate, fimbriate. Male flowers: bracteoles c. 1.5 mm long, laciniate; stamens c. 3 mm long. Female flower: styles to 4 mm long, united for 1 mm, apices spreading, bifid. Capsule and seeds not seen.

Mozambique. N: 9.5 km east of Nampula, st. 23.vii.1962, *Leach & Schelpe* 11436 (K; LISC; SRGH).
Known only from Niassa Province. Granite slopes; 400–500 m.

119. **Euphorbia unicornis** R.A. Dyer in Bothalia **6**: 225 (1951). —Court, Succ. Fl. South. Africa: 28 (1981). Type: Mozambique, Meluco Distr., Cuero Mte., fl. 14.ii.1949 *Pedro & Pedrógão* 5091 (PRE, holotype).

Spiny succulent shrublet to c. 30 cm high, branching from the base. Branches 10 mm in diameter, terete, with 6–7 angles or ridges separated by grooves; angles with prominent tubercles c. 1.5 mm high and 5–7 mm apart along the margins. Spine shields joined and completely covering the tubercles, forming continuous, horny, whitish ridges, c. 4 mm wide, separated by pale green grooves 1 mm wide; spines single, 4–6 mm long; prickles 1–1.5 mm long. Leaves c. 1 mm long, deciduous. Cymes solitary, simple; peduncles and cyme branches 1–2 mm long; bracts 1 mm long, oblong. Cyathia c. 2 × 3 mm, with cup-shaped involucres; glands c. 1.5 mm wide, touching to form a circle, red; lobes 1 × 1 mm, subquadrate, denticulate. Male flowers: bracteoles c. 1.5 mm long, laciniate; stamens c. 3 mm long. Female flower: ovary sessile; styles 2 mm long, united to halfway, apices spreading, ± bifid. Capsules and seeds not seen.

Mozambique. N: Meluco Distr., Cuero Mte., fl. 14.ii.1949, *Pedro & Pedrógão* 5091 (PRE).
Known only from the type collection. Rocky outcrops; 450–740 m.

60. SYNADENIUM Boiss.

Synadenium Boiss. in de Candolle, Prodr. **15**, 2: 187 (1862). —N.E. Brown in F.T.A.
6, 1: 462 (1911). —S. Carter in F.T.E.A., Euphorbiaceae, part 2: 534 (1988). —
Radcliffe-Smith, Gen. Euphorbiacearum: 417 (2001).

Shrubs or trees with copious caustic milky latex, monoecious.   Branches
cylindrical, ± fleshy and marked with large elliptic leaf scars. Leaves fleshy, with
stipules modified as small but ± conspicuous dark brown glands.   Inflorescence
with sessile cyathia in dichotomously branching axillary cymes, usually crowded
into pseudo-umbels (pseudumbels) at the branch tips; bracts paired, persistent.
Cyathia with numerous male flowers in 5 groups surrounding a solitary female
flower and all enclosed within a cup-like involucre. Involucres with an entire or
occasionally notched spreading furrowed glandular rim surrounding 5 fringed
lobes. Male flowers in 5 groups, with stamens shortly exserted from the involucre,
bracteoles included. Female flower shortly pedicellate, the pedicel elongating
slightly in fruit; perianth reduced to a 3-lobed rim below the ovary; styles 3, joined
at the base, with bifid stigmas. Capsule 3-lobed, dehiscent. Seeds with a sessile,
often rudimentary caruncle.

A genus confined to east and southern tropical Africa, with 14 closely related species, of
which 6 occur in the Flora Zambesiaca area.
Differences between the species, at least from herbarium specimens, appear to be slight, and
have usually been based upon leaf and cyme-branching characters. However, emphasis should
also be placed upon habit, shape and size of the cyathium, colour and furrowing of the
involucral glands, and features of the capsules and seeds. Relatively little fruiting material has
been collected and details are missing for the majority of species in the Flora Zambesiaca area.

1. Glandular rim of involucre and the involucre itself with a deep cleft on one side · · · · · 2
– Glandular rim entire around the involucre · 4
2. Leaf margins entire · 2. *glabratum*
– Leaf margins minutely serrate · 3
3. Inflorescences lax on peduncles to 5 cm long · 1. *angolense*
– Inflorescences compact on peduncles to 3 cm long · · · · · · · · · · · · · · · · · · 3. *halipedicola*
4. Leaves completely glabrous; midrib sharply keeled on lower surface with margin distinctly
   crenulate · 4. *cupulare*
– Leaves glabrous or hairy; midrib rounded, or if keeled, not distinctly crenulate · · · · · · 5
5. Leaves glabrous on both surfaces, rarely sparsely pubescent on the margins and the midrib
   beneath towards the base · 5. *cameronii*
– Leaf upper surfaces distinctly pubescent, at least along the margins, the midrib beneath
   always pubescent · 6. *kirkii*

1. **Synadenium angolense** N.E. Br. in F.T.A. **6**, 1: 469 (1911). TAB. **85**, fig. A. Type from Angola.
     Synadenium grantii sensu N.E. Brown in F.T.A. **6**, 1: 468 (1911). —sensu F. White,
   F.F.N.R.: 204 (1962), in note on Worthington s.n. —sensu Topham, Check List For. Trees
   Shrubs Nyasaland Prot.: 53 (1958), non Hook.f.
     Synadenium kirkii sensu Hargreaves, Succ. Spurges Malawi: 79 (1987) pro parte as to
   distr. Malawi, Northern Region, non N.E. Br.

Shrub to 3 m, or small tree to 6 m high, branches and stems fleshy. Leaves to
12.5–17 × 6.5 cm, oblanceolate to obovate, tapering at the base into a winged petiole
to c. 1 cm long, ± acute at the apex, entire (or minutely serrate), glabrous but a few
hairs sometimes present on the basal margins and the midrib beneath; midrib very
prominent and rounded below. Cymes 3–5, axillary, pseudumbellate, on peduncles
2–5 cm long, each 2–4 times dichotomously forked with a sessile cyathium
(involucre) in each fork and at the tips of the ultimate branchlets; cyme branches
1.5–2 cm long, green or flushed purplish, thinly and minutely pubescent, otherwise
glabrous; bracts c. 4 × 4 mm, subquadrate, denticulate, sparsely pilose. Cyathia c. 3
× 5.5 mm, with broadly funnel-shaped involucres, deeply notched on one side,
pubescent around the base; glandular rim c. 1 mm wide, furrowed; lobes
subquadrate, c. 1.75 × 1.75 mm, denticulate. Male flowers: bracteoles 2 mm long,
filamentous, plumose; stamens 3.5 mm long. Female flower: perianth of 3 linear

Tab. 85. A. —SYNADENIUM ANGOLENSE. A1, apical portion of flowering branch (× 2/3),
from *Leach & Williamson* 13546; A2, cyme (× 3), from *Gossweiler* 14078 (spec. from Angola).
B. —SYNADENIUM HALIPEDICOLA. B1, apical portion of flowering branch (× 2/3);
B2, cyme (× 3), B1 & B2 from *Leach* 11262. Drawn by Eleanor Catherine.

lobes 1.5–2 mm long (or acutely 3-lobed); ovary densely pubescent; styles c. 2 mm long, joined for one-third, deeply bifid. Capsule 6 × 5.5 mm, deeply acutely lobed, glabrescent; pedicel c. 2 mm long. Seeds 2.4 × 2 mm, ovoid, slightly compressed, minutely verrucose, grey; caruncle 0.4 mm in diameter.

Zambia. N: Nchelenge Distr., Lake Mweru, fl. & fr. 6.viii.1958, *Fanshawe* 4658 (K; FHO; NDO). W: Mwinilunga Distr., Ikelenge Village, c. 13 km south of Kalene Hill, fl. 28.x.1966, *Leach & Williamson* 13546 (K; SRGH). **Malawi**. N: Karonga, Mbande Hill, fl. & fr. 29.vii.1978, *Hargreaves* 474 (MAL).
Also in western Angola and southeastern Dem. Rep. Congo. Mixed deciduous woodland and dense thickets; 550–1400 m.
Leach & Williamson recorded that it is used as a hedge in Ikelenge Village, while A.M. Fleming recorded (ex Mr. Worthington) that it is sought after as a cure for leprosy by people of Barotseland, who travel many miles to the Zambezi Valley to get it.
Specimens from the Northern Provinces of Zambia and Malawi possess leaves with minutely serrate margins instead of entire margins typical of the species. The perianth lobes of the female flower are also no more than acute instead of produced into linear filaments. Further collections may prove the existence of a distinct taxon.

2. **Synadenium glabratum** S. Carter in Kew Bull. **42**: 667 (1987). Type: Zambia, Lake Tanganyika, Crocodile Island, fl. 12.iv.1959, *Richards* 11209 (K, holotype).

Shrub c. 1.5 m high, with branches and stems fleshy. Leaves to c. 15 × 5 cm, linear-lanceolate to oblanceolate, tapering at the base to a petiole to 1 cm long, obtuse at the apex, entire, glabrous; midrib acutely prominent beneath. Cymes 3–5, axillary, pseudumbellate, on glabrous peduncles to 4 cm long, each 2 times dichotomously forked with a sessile cyathium (involucre) in each fork and at the tips of the ultimate branchlets; cyme branches 5–15 mm long, glabrous or rarely the uppermost shortly pubescent; bracts c. 3 × 3 mm, subquadrate, dentate, glabrous or sparsely pilose. Cyathia c. 2 × 4 mm, with shallowly cup-shaped involucres, deeply notched on one side, glabrous or shortly puberulous at the base; glandular rim c. 0.8 mm wide, furrowed, yellow becoming reddish; lobes c. 1.5 × 1.5 mm, rounded, glabrous. Male flowers: bracteoles c. 1.5 mm long, laciniate, plumose; stamens 2.8 mm long. Female flower: ovary (only immature seen) pubescent; perianth acutely 3-lobed. Capsule and seeds not seen. (Capsule and seed data included in the *S. glabratum* protologue were taken from *Richards* 19104, identified in error as this species).

Zambia. N: Mbala Distr., Lake Tanganyika, Crocodile Island, fl. 9.ii.1964, *Richards* 18968 (K). Known only from Crocodile Island. In sand and pebbles on rocky shore; 775 m.

3. **Synadenium halipedicola** L.C. Leach in Garcia de Orta, Sér. Bot. **6**: 47 (1983). TAB. **85**, fig. B. Type: Mozambique, Nhamatanda Distr., 29 km south of Muda on Nhamatanda/Búzi district boundary, collected June 1959, flowered in Harare 14.v.1975, *Leach* 11262 (LISC, holotype; BM; BR; K; MO; NBG; NH; PRE; SRGH).

Shrub to 3 m high, with branches and stems fleshy. Leaves to 15.5 × 7 cm, obovate, shortly petiolate, acute at the apex, minutely serrate, glabrous; midrib sharply prominent. Cymes 2–5, axillary, pseudumbellate, on sparsely hairy peduncles 2–3 cm long, green becoming purplish, each 1–2 times dichotomously forked with a sessile cyathium (involucre) in each fork and at the tips of the ultimate branchlets; cyme branches 5–10 mm long, pubescent; bracts c. 3 × 4 mm, subquadrate, denticulate, sparsely hairy. Cyathia c. 3 × 6.5 mm, with broadly cup-shaped involucres, deeply notched on one side, sparsely pubescent towards the base; glandular rim 1 mm wide, deeply and densely furrowed; lobes c. 2.5 × 2.5 mm, subquadrate, denticulate, sparsely pilose. Male flowers: bracteoles c. 3 mm long, filamentous, plumose; stamens 3.5 mm long. Female flower: perianth reduced to a 3-lobed rim; styles c. 1.5 mm long, joined at the base, bifid to halfway. Capsule c. 9 × 10 mm, obovoid, pedicel c. 2 mm long, pubescent. Seeds c. 3 × 2.75 mm, ovoid, minutely verrucose; caruncle minute.

Mozambique. MS: Nhamatanda Distr., 29 km south of Muda on Nhamatanda/Búzi district boundary, collected June 1959, cultivated and flowered in Harare 14.v.1975, *Leach* 11262 (BM; BR; K; LISC; MO; NBG; NH; PRE; SRGH).
Known only from the type collection. Mixed deciduous woodland; c. 75 m.

4. **Synadenium cupulare** (Boiss.) Wheeler in White, Dyer & Sloane, Succ. Euphorb. **2**: 953 (1941). —Coates-Palgrave, Trees Southern Africa: 453 (1977). Type from South Africa (KwaZulu-Natal).
  Euphorbia cupularis Boiss., Cent. Euph.: 23 (1860).
  Synadenium arborescens E. Mey. ex Boiss. in de Candolle, Prodr. **15**, 2: 187 (1862). —N.E. Br. in F.C. **5**, 2: 221 (1915). Type as for *S. cupulare*.

Shrub to 2 m or tree to 4 m high, with branches and stems fleshy. Leaves entirely glabrous, to 14 × 5.5 cm, obovate, obtuse at the apex, tapering at the base into a petiole 5–10 mm long, margins entire, lower surface often blotched with red and midrib with a crenulate keel 1 mm wide. Cymes 2–5, axillary, pseudumbellate, on sparsely pubescent peduncles to 3 cm long, each usually 2 times dichotomously forked with a sessile cyathium (involucre) in each fork and at the tips of the ultimate branchlets; cyme branches to 1 cm long, pubescent, otherwise glabrous; bracts c. 3 × 3 mm, subquadrate, irregularly toothed, sparsely hairy. Cyathia c. 3 × 5.5 mm, with cup-shaped involucres, sparsely hairy at the base; glandular rim c. 0.75 mm wide, distinctly furrowed; lobes c. 1.5 × 1.5 mm, subquadrate, denticulate. Male flowers: bracteoles 2 mm long, filamentous, plumose; stamens 3.5 mm long. Female flower: ovary pubescent; styles 2 mm long, joined for one-third, apices bifid. Capsule and seeds not seen.

Mozambique. GI: Xai-Xai Distr., Chiconela, fl. 18.v.1983, *Calane da Silva, Jansen, Marime & Manhiça* 175 (WAG).
Also in South Africa and Swaziland. Open woodland; c. 38 m.

5. **Synadenium cameronii** N.E. Br. in Bull. Misc. Inform., Kew **1901**: 133 (1901); in F.T.A. **6**, 1: 463 (1911). —Hargreaves, Succ. Spurges Malawi: 78 (1987) excl. spec. *Williams* 188. Type: Malawi, Namadzi (Namasi) R., fl. iv.1899, *Cameron* s.n. (K, holotype).
  Synadenium gazense sensu Drummond in Kirkia **10**: 253 (1975) non N.E. Br.
  Synadenium grantii sensu Hargreaves, Succ. Spurges Malawi: 77 (1987) excl. desc. non Hook.f.
  Synadenium sp. aff. grantii sensu Coates Palgrave, Trees Southern Africa: 454 (1977).

Shrub to 3 m or small tree to 5 m high, with branches and stems fleshy. Leaves to c. 18 × 8 cm, oblanceolate, apiculate at the apex, tapering at the base into a winged petiole to 1 cm long, margins minutely crisped, both surfaces glabrous but a very few sparse hairs occasionally present on the margins and the midrib beneath at the base; midrib beneath with a sharp keel to 1 mm wide. Cymes 3–5, axillary, pseudumbellate, on sparsely hairy peduncles 2–5 cm long, each 1–2 times dichotomously forked with a sessile cyathium (involucre) in each fork and at the tips of the ultimate branchlets; cyme branches c. 1–1.5 cm long, pubescent; bracts c. 3 × 3 mm, subquadrate, dentate, sparsely pubescent. Cyathia c. 3.5 × 6 mm, with funnel-shaped involucres, pubescent around the base; glandular rim c. 1.25 mm wide, furrowed; lobes c. 1.5 × 2 mm, denticulate, very sparsely hairy. Male flowers: bracteoles 2.5 mm long, filamentous, plumose; stamens 3.25 mm long. Female flower: ovary pubescent; styles 2 mm long, joined at the base, deeply bifid. Capsule 7 × 6 mm, obovoid, deeply acutely lobed, apex depressed. Seeds 2.5 × 1.75 mm, ovoid, minutely verrucose, grey; caruncle minute.

Zimbabwe. N: Makonde Distr., near Chinhoyi (Sinoia), by Manyame (Hunyani) R., fl. 22.iv.1948, *Rodin* 4378 (K; WAG). W: Insiza Distr., 24 km east of Filabusi, fl. 24.iv.1954, *Plowes* 1724 (K; SRGH). C: Harare (Salisbury), fl. (no date), *Christian* 378 (K). E: Nyanga (Inyanga), near Cheshire, fl. 4.ii.1931, *Norlindh & Weimarck* 4780 (K). S: Masvingo Distr., Mutirikwi Recreational Park (Kyle National Park), fl. & fr. 24.v.1971, *Grosvenor* 542 (K; LISC). **Malawi**. S: Chikwawa Distr., Kapichira (Livingstone Falls), fl. 21.iv.1970, *Brummitt* 9996 (K; MAL). **Mozambique**. T: Songo, near Mucangádzi R., fl. 7.iii.1972, *Macêdo* 5016 (LISC). MS: 20 km south of Chimoio (Vila Pery), fl. viii.1956, *Leach & Pienaar* 5111 (BR; K).
Not known elsewhere. Rocky outcrops and exposed granite slopes, often also beside rivers, in mixed deciduous woodland; 100–1400 m.

6. **Synadenium kirkii** N.E. Br. in F.T.A. **6**, 1: 466 (1911). —Hargreaves, Succ. Spurges Malawi: 79 (1987), excl. distr. in N Region. —S. Carter in Kew Bull. **55**: 441 (2000). Type: Malawi, near Sekwene Village [northeast of Chikwawa], fl. iv.1859, *Kirk* s.n. (K, holotype).

Synadenium gazense N.E. Br. in F.T.A. **6**, 1: 467 (1911). —S. Moore in J. Linn. Soc., Bot. **40**: 190 (1911). —Eyles in Trans. Roy. Soc. South Africa **5**: 400 (1916) as *"gazenae"*. Type: Zimbabwe, Chipudzana (Chipetzana) R., fl. 19.iv.1907, *Swynnerton* 1505 (BM, holotype).
   Synadenium grantii sensu N.E. Brown in F.T.A. **6**, 1: 468 (1911) pro parte as to *Menyharth* 614 non Hook.f.

Shrubs 2–3 m high, or small trees to c. 4 m high, with branches and stems fleshy. Leaves to 11–18 × 5–8 cm, obovate to broadly obovate, obtuse to shortly apiculate at the apex, tapering to a subsessile base, pubescent or thinly so on upper surface of mature leaves at least on the lower third and along the margins; midrib keeled below, pilose or densely pubescent.   Cymes 2–6, axillary, pseudumbellate, on pubescent peduncles 2.5–4 cm long, each 1–2 times dichotomously forked with a sessile cyathium (involucre) in each fork and at the tips of the ultimate branchlets; cyme branches 1–2 cm long, pubescent; bracts c. 4 × 4 mm, subquadrate, dentate, thinly hairy. Cyathia c. 3.5 × 6 mm, with funnel-shaped involucres, pubescent on the lower half; glandular rim c. 0.5 mm wide, furrowed, yellow; lobes c. 2 × 2 mm, rounded, denticulate. Male flowers: bracteoles c. 2 mm long, plumose; stamens c. 3 mm long. Female flower: ovary pubescent; styles c. 1.75 mm long, joined for one-third, apices bifid. Capsule 6.5 × 6 mm, deeply and acutely lobed, thinly pubescent; pedicel c. 2 mm long. Seeds 2.8 × 2.2 mm, ovoid, densely and minutely verrucose, yellowish-buff; caruncle minute.

Zimbabwe. N: Mutoko Distr., Mudzi Dam, fl. 16.ii.1962, *Wild* 5676 (K; SRGH). C: Harare (Salisbury), fl. vii.1923, *Dept. Agric.* 3295 (K). E: Chirinda, fl. 11.v.1943, *W. Williams* in *GHS* 10001 (K). **Malawi**. N: Nkhata Bay Distr., Likoma Island, Chindandari Hills, fr. 20.viii.1984, *Salubeni & Nachamba* 3842 (K; MAL). C: Lilongwe to Dedza road, near Nathenje, fl. 30.iv.1989, *Radcliffe-Smith, Pope & Goyder* 5802 (K). S: Thyolo Distr., Zoa Falls near Thekerani, fl. 18.iv.1982, *Chapman* 6132 (BR; K; FHO; MAL). **Mozambique**. Z: Morrumbala Distr., Mbobo, fl. 20.v.1943, *Torre* 5342 (LISC). T: Mutarara Distr., 35.8 km from Mutarara-a-Velha towards Sinjal, fr. 18.vi.1949, *Barbosa & Carvalho* 3141 (K; LISC). MS: Guro Distr., 6 km from Mungári on Nhacolo (Tambara) road, fl. 12.v.1971, *Torre & Correia* 18386 (LISC).
   Not known elsewhere. Rocky outcrops, rocky slopes of escarpments and gorges, open mixed deciduous woodland; 45–1000 m.

61. MONADENIUM Pax

Monadenium Pax in Bot. Jahrb. Syst. **19**: 126 (1894). —N.E. Brown in F.T.A. **6**, 1: 450 (1911). —Bally, Genus Monadenium: 14 (1961). —S. Carter in F.T.E.A., Euphorbiaceae, part 2: 540–564 (1988). —Webster in Ann. Missouri Bot. Gard.: 81 (1994). —Radcliffe-Smith, Gen. Euphorbiacearum: 417 (2001).
   Lortia Rendle in J. Bot. **36**: 29 (1898).
Stenadenium Pax in Bot. Jahrb. Syst. **30**: 343 (1901). —N.E. Brown in F.T.A. **6**, 1: 448 (1911). —Brenan, Check-list For. Trees Shrubs Tang. Terr.: 227 (1949).

Small trees, shrubs or perennial herbs, often geophytic (with aerial parts dying back after the growing season) in the Flora Zambesiaca area, with fleshy or succulent stems and branches and with caustic milky latex, monoecious; roots thick and fleshy, often tuberous. Leaves sometimes fleshy, glabrous or hairy, with stipules apparently absent, or occasionally modified as glands or spines.   Cymes axillary, with sessile cyathia branching dichotomously, or cyathia rarely solitary; bracts persistent, paired, free, or partly united along the ventral margin to form an incomplete bract-cup and enveloping the involucre.   Cyathia with numerous male flowers in 5 groups surrounding a solitary female flower and all enclosed within a cup-like involucre. Involucres with glands fused to form an entire rim deeply notched on the dorsal side, and surrounding the 5 fringed lobes. Male flowers in 5 groups, bracteolate, scarcely exserted.   Female flower pedicellate; perianth reduced to a rim below the ovary, rarely shortly 3-lobed; styles 3, joined at the base, stigmas bifid. Capsule exserted through the notch in the glandular rim, dehiscent. Seeds with or without a caruncle.

A well-defined genus with currently over 70 species recognised, of which 20 occur in the Flora Zambesiaca area. Distribution is generally sparse, but extends throughout the eastern tropical regions of Africa, from northern Somalia southwards to the Northern Province of South Africa and westwards into Angola.

The genus is distinguished primarily by the entire, horseshoe-shaped involucral gland, which has a wide rim extended to protect the ovary exserted through the notch. The persistent bracts, which envelope the involucre, are usually united behind the glandular rim and are sometimes large and showy. The seeds are usually oblong, with a relatively large mushroom-shaped caruncle capping the apex, or sometimes ovoid in some geophytic species and either with or without a rounded caruncle. In the Flora Zambesiaca area the habit ranges from small geophytes to herbs or shrubs with stems ± fleshy, or occasionally strongly succulent.

1. Succulent perennials with stems to c. 3 cm in diameter, tessellated (with a diamond-shaped pattern of tubercles) · 2
– Shrubs or perennial herbs, often geophytic, stems not tessellated · · · · · · · · · · · · · · 3
2. Tessellations with shallow tubercles (see TAB. **88**, fig. B1); leaf scars without prickles · 19. *lugardiae*
– Tessellations with prominent tubercles c. 5 mm high; leaf scars with a cluster of 5–7 prickles around the base · 20. *spinulosum*
3. Shrub or small tree to 3 m high; leaf scar with a stout spine at the base · · · · · · · 18. *torrei*
– Geophytes or perennial herbs to c. 80 cm high; leaf scars without spines · · · · · · · · · · · 4
4. Cyathial bracts completely free · 5
– Cyathial bracts around the involucre united ventrally, at least at the base · · · · · · · · · · 11
5. Flowering stems appearing before the leaves · · · · · · · · · · · · · · · · · · 3. *orobanchoides*
– Flowers appearing with the leaves · 6
6. Leaves linear, c. 3 mm wide, clustered at stem apex · · · · · · · · · · · · · · · · · · 2. *filiforme*
– Leaves linear-lanceolate to obovate, more than 5 mm wide · · · · · · · · · · · · · · · · · · 7
7. Cymes on peduncles up to 8 cm long, or cyathia solitary; seeds ovoid without a caruncle · · 8
– Cymes on peduncles less than 2 cm long, cyathia rarely solitary; seeds oblong, 4-angled, with a caruncle · 9
8. Leaves on an elongating stem, linear-lanceolate to obovate · · · · · · · · · · 1. *pedunculatum*
– Leaves in a rosette, spreading (see TAB. **86**, fig. A1) · · · · · · · · · · · · · · · · · 4. *nervosum*
9. Leaves in a rosette, spreading; lamina obovate, c. 3.5 × 2.5 cm; cymes simple or reduced to solitary cyathia · 8. *mapingense*
– Leaves not in a rosette, on a stem to 20 cm high or more; lamina oblanceolate; cymes 1–2-forked · 10
10. Leaf margins with small distant teeth; cymes subsessile · · · · · · · · · · · · · · · 9. *capitatum*
– Leaf margins entire; cymes on peduncles 0.5–2 cm long · · · · · · · · · · · · · 10. *herbaceum*
11. Plant c. 1 cm high; leaves in a basal rosette; seeds without a caruncle · · · · 5. *pudibundum*
– Plant more than 5 cm high; leaves not forming a rosette; seeds with a caruncle · · · · · · 12
12. Seeds ovoid or subglobose, with a rounded caruncle · 13
– Seeds oblong, 4-angled, with a cap-like caruncle (see TAB. **87**, fig. 5) · · · · · · · · · · · 14
13. Plant glabrous; bract-cup ± equalling the involucre · · · · · · · · · · · · 6. *pseudoracemosum*
– Whole plant covered with short bristles; bract-cup nodding, completely enveloping the involucre · 7. *discoideum*
14. Cymes on peduncles up to 2 cm long, or subsessile · 15
– Cymes on peduncles 5–10 cm long · 18
15. Leaves broadly obovate, less than twice as long as broad, surface distinctly roughened; plants up to 50 cm high or more · 11. *rugosum*
– Leaves lanceolate to obovate, more than twice as long as broad, surface not rough · · · 16
16. Plants 30–60 cm high; leaves obovate, hairy; bract-cup notched to halfway between rounded lobes · 14. *parviflorum*
– Plants less than 20 cm high; leaves narrowly lanceolate to obovate, glabrous; bract-cup deeply notched between acute lobes · 17
17. Cymes on peduncles c. 3 mm long; bract-cup suberect, notched almost to the base between very acute lobes · 12. *fwambense*
– Cymes on peduncles up to 10 mm long; bract-cup nodding, notched to halfway or more between ± acute lobes · 13. *crenatum*
18. Leaves densely pilose hairy, with hairs c. 1 mm long · · · · · · · · · · · · · · · · · 15. *hirsutum*
– Leaves ± sparsely hispid with stiff or cartilaginous hairs · 19
19. Plant up to 80 cm high; leaves with stiff hairs; peduncles glabrous · · · · · · · · · · · 16. *laeve*
– Plant up to 45 cm high; leaves with cartilaginous hairs; peduncles with scattered cartilaginous prickles · 17. *friesii*

1. **Monadenium pedunculatum** S. Carter in Kew Bull. **42**: 903 (1987); in F.T.E.A.,
Euphorbiaceae, part 2: 542, fig. 102/1–3 (1988). Type from Tanzania.
   Monadenium chevalieri sensu P.R.O. Bally, Gen. Mon.: 30 (1961) pro parte, as to Zambian
distribution. —sensu Cribb & Leedal, Mount. Fl. S. Tan.: 78 (1982), non N.E. Br.

Geophyte; rootstock tuberous, up to 6 cm in diameter, napiform, producing 1–2
woody stems to c. 6 cm long below ground. Annual (aerial) stems 1–4, usually
simple, 2–15(20) cm high, sometimes minutely papillose. Leaves sessile, variable,
from up to 9 × 0.5 cm and linear, or up to 6 × 1.5 cm and lanceolate, or up to 4 × 2
cm and obovate; stipules glandular, minute, evident on young growth only. Cymes
on peduncles up to 5(8) cm long, simple with cyme branches to 1 cm long, or cyathia
solitary; bracts free, shorter than the involucre, c. 3 × 2 mm, oblong, margin
denticulate. Cyathia c. 5 × 3 mm, with barrel-shaped involucres; glandular rim 1.5–2
mm high, entire to crenulate, white or pink; lobes c. 1 × 1 mm, rounded. Male
flowers: bracteoles few, filamentous; stamens 3.2 mm long. Female flower: styles 1
mm long, with thickened rugulose bifid apices. Capsule c. 5 × 5.5 mm, obtusely 3-
lobed, exserted on a reflexed pedicel 6–12 mm long, smooth. Seeds c. 2.8 × 2.3 mm,
conical with truncate base, minutely and densely verrucose, brown becoming
purplish; caruncle absent.

   Zambia. N: Mporokoso, fl. 7.x.1958, *Fanshawe* 4888 (K; NDO); 100 km east of Kasama, fl. &
fr. 26.xi.1960, *E.A. Robinson* 4121 (K). **Malawi**. N: Chitipa Distr., Kaseye to Chibula, fl.
21.xii.1978, *Hargreaves* 600 (MAL). **Mozambique**. N: Ribáuè Distr., Matamane, fl. 5.xi.1942,
Mendonça 1253 (LISC).
   Also in Burundi, southeast Dem. Rep. Congo, and in southwest Tanzania where it is
widespread. Sandy soil in open *Brachystegia* woodland; 450–1525 m.

2. **Monadenium filiforme** (P.R.O. Bally) S. Carter in Kew Bull. **42**: 905 (1987). Type: Zambia,
Kawambwa Distr., near Falls, fl. & fr. 13.xi.1957, *Fanshawe* 4001 (K, holotype; BR; NDO).
   Monadenium chevalieri var. *filiforme* P.R.O. Bally, Gen. Mon.: 32 (1961).

Geophyte; rootstock tuberous, c. 2.5 × 2 cm, napiform, usually producing a single
fleshy stem below ground up to 4 cm long and 3.5 mm in diameter. Leaves on lower
stem scale-like, to 16 × 2.5 mm, lanceolate; leaves on upper part of the stem crowded,
erect, to 7.5 cm long, 3 mm wide, linear. Cymes solitary on peduncles to 3 cm long,
simple, with cyme branches to 4 mm long; bracts free, longer than the involucre, 3–6
mm long, linear. Cyathia 4 × 2.5 mm, with broadly funnel-shaped involucres;
glandular rim c. 2 mm high, spreading to 4 mm wide, margin crenulate, white or
pink; lobes 1 × 1 mm, subquadrate, denticulate. Male flowers: bracteoles few,
filiform; stamens c. 3.5 mm long. Female flower: styles 1.5 mm long, stigmas bifid.
Capsule 4.5 × 3.5 mm, obtusely lobed, smooth; exserted on a reflexed pedicel to 6
mm long. Seeds 3 × 2 mm, ovoid, smooth; caruncle absent.

   Zambia. N: Kawambwa Distr., near Falls, fl. & fr. 13.xi.1957, *Fanshawe* 4001 (BR; K; NDO).
   Known only from the type and from one other collection from neighbouring Dem. Rep.
Congo (Kundelungu). Miombo woodland and wooded grassland, by falls; c. 1400 m.

3. **Monadenium orobanchoides** P.R.O. Bally in Candollea **17**: 30 (1959); Gen. Mon.: 36 (1961).
—Binns, First Check List Herb. Fl. Malawi: 51 (1968). —S. Carter in F.T.E.A.,
Euphorbiaceae, part 2: 544 (1988). Type from Tanzania.
   Monadenium sp. nov. of Brenan in Mem. New York Bot. Gard. **9**: 67 (1954).
   Monadenium chevalieri var. *spathulatum* P.R.O. Bally, Gen. Mon.: 33 (1961). —Binns, First
Check List Herb. Fl. Malawi: 51 (1968). —Hargreaves, Succ. Spurges Malawi: 64 (1987).
Type: Malawi, Nyika Plateau, 3 km southwest of Rest House, fl. 21.x.1958, *Robson & Angus*
224 (K, holotype).
   Monadenium chevalieri sensu Hargreaves, Succ. Spurges Malawi: 63 (1987) non N.E. Br.
excl. desc.
   Monadenium chevalieri var. *filiforme* sensu Hargreaves, Succ. Spurges Malawi: 65 (1987),
non P.R.O. Bally.

Geophyte; rootstock tuberous, up to c. 5 cm in diameter, napiform, producing 1–4
stems up to 4 cm long below ground. Annual (aerial) stems smooth, or minutely
papillose, with leaves borne in rosettes at apices of stems to 2 cm high; flowering
stems up to 6 cm high, the bases with crowded scale-like leaves to 10 × 5 mm, oblong.

Leaves appearing after the flowers, sessile, up to 6 × 2.5 cm, obovate, lower surface often reddish; stipules apparently absent. Cymes crowded at stem apex on peduncles 0.5–2.5(3.5) cm long, 1–2-forked, or cyathia solitary; bracts free, 2–4 × 2 mm, oblong, about equal to the involucres, apices denticulate. Cyathia c. 3 × 2.5 mm, with barrel-shaped involucres; glandular rim c. 1.2 mm high, undulate, white or pink; lobes c. 0.5 × 0.5 mm, subquadrate. Male flowers: bracteoles few, 1.5 mm long, filamentous; stamens 2 mm long. Female flower: styles 1 mm long, stigmas bifid, spreading, thickened. Capsule c. 3.5 × 4 mm, obtusely 3-lobed with 2 minute fleshy crenulate ridges along each angle, exserted on a reflexed pedicel 3–4 mm long. Seeds 2 × 1 mm, ovoid with truncate base, pale brown, areolate; caruncle absent.

Zambia. N: Mbala Distr., Uningi Pans, fl. 23.x.1966, *Richards* 21552 (K). **Malawi**. N: Mzimba Distr., 6.5 km southwest of Chikangawa, fl. & fr. 14.ix.1978, *E. Phillips* 3926B (K; MAL). C: Ntchisi Distr., Chinthembwe (Chintembwe), fl. & fr. 9.ix.1946, *Brass* 17585 (K; MAL).
Also in southern Tanzania. Montane grassland and wooded grassland, often amongst rocks, appearing after burning; 1500–2300 m.

4. **Monadenium nervosum** P.R.O. Bally in Candollea 17: 29 (1959); Gen. Mon.: 22 (1961). —S. Carter in F.T.E.A., Euphorbiaceae, part 2: 544 (1988). TAB. **86**, fig. A. Type: Zambia, Mbala Distr., Mpulungu, beside Lake Tanganyika, fl. & fr. 16.xii.1954, *Richards* 3656 (K, holotype).
   Monadenium letouzeyanum Malaisse in Bull. Mus. Natl. Hist. Nat., sér. 4, **11**, sect. B, Adansonia No. 4: 337 (1989). Type from Dem. Rep. Congo.

Geophyte; rootstock tuberous, up to 7 cm in diameter, napiform, producing 1–3 stems from 1–3 cm long below ground. Annual (aerial) stems 2–6 cm high. Leaves scale-like and c. 4 × 2 mm on lower stems, with upper leaves clustered in a rosette at the stem apex, these up to 9 × 4.5 cm, obovate and often flecked with red; petiole to 1 cm long; stipules apparently absent. Cymes reduced to solitary cyathia on minutely papillose peduncles to 6 cm long, sometimes appearing before the leaves; bracts free, c. 2.5 × 2.5 mm, subquadrate, shorter than the involucres, denticulate. Cyathia c. 4.5 × 3 mm, with barrel-shaped involucres; glandular rim c. 1.5 mm high, undulate, pink or red; lobes 1 × 1 mm, rounded, denticulate. Male flowers: bracteoles few, 1.5 mm long, laciniate; stamens 2 mm long. Female flower: styles 1 mm long, with thickened deeply bifid apices. Capsule c. 4 × 4.5 mm, obtusely 3-lobed, smooth, exserted on a reflexed pedicel 5–7 mm long. Seeds ovoid with truncate base, 2.3 × 1.3 mm, greenish-brown, minutely tuberculate; caruncle absent.

Zambia. N: Mbala Distr., Mpulungu, fl. & fr. 17.xi.1959, *Richards* 11780 (K); Kawambwa Distr., M'bereshi Furrow, fr. 30.xi.1961, *Richards* 15458 (K).
Also in southwestern Tanzania (Ufipa Distr.) and southeastern Dem. Rep. Congo (Katanga Province). *Brachystegia* woodland on stony hillsides and lateritic rocky places, in gritty soils with short grasses; 750–1420 m.

5. **Monadenium pudibundum** P.R.O. Bally in Candollea 17: 31 (1959). Type: Zambia, Mwinilunga Distr., c. 0.8 km south of Matonchi Farm, fl. & fr. 30.x.1937, *Milne-Redhead* 3011 (K, holotype; BR).
   Monadenium simplex var. *pudibundum* (P.R.O. Bally) P.R.O. Bally, Gen. Mon.: 40 (1961).

Geophyte; rootstock tuberous, 5–6 × 2–3 cm, napiform, producing a stem c. 1 cm long below ground. Leaves appearing after the flowers, produced in a rosette at ground level, c. 3.2 × 1.5 cm, obovate; flowering stems with scale-like leaves. Cymes 1–5, reduced to solitary cyathia on peduncles 1–2.5 cm long; bracts joined in a bract-cup 5 × 8 mm, longer than the involucre, shortly notched between rounded apices, whitish. Cyathia c. 5 × 4 mm, with barrel-shaped involucres; glandular rim c. 1.5 mm high, crenulate, white; lobes 1 × 1 mm, subquadrate, denticulate. Male flowers: bracteoles 1.5 mm long, puberulous; stamens c. 2 mm long. Female flower: perianth obvious, 3-lobed, toothed; styles 1 mm long, with bifid thickened apices. Capsule 4 × 3.5 mm, obtusely 3-lobed, smooth. Seeds 2 × 1.25 mm, ovoid, smooth; caruncle absent.

Leaf upper surface glabrous · var. *pudibundum*
Leaf upper surface densely hairy · var. *lanatum*

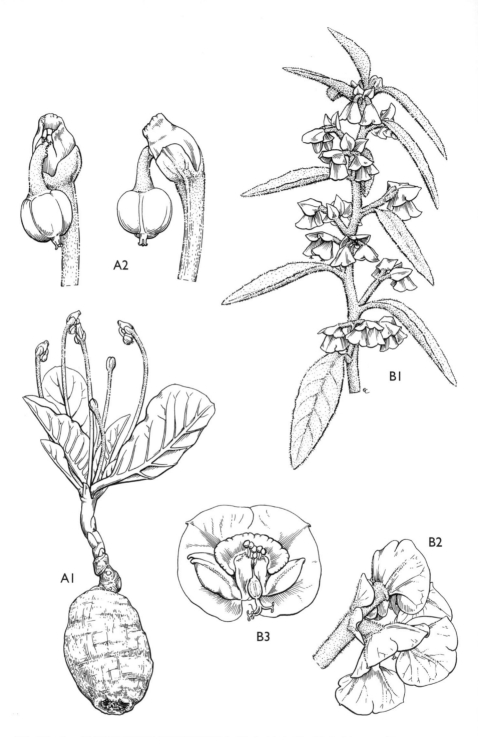

Tab. 86. A. —MONADENIUM NERVOSUM. A1, habit (× 1); A2, fruiting cyathia (× 4), A1 &
A2 from *Richards* 3685. B. —MONADENIUM HIRSUTUM. B1, habit (× 1), from a
photo; B2, cyme (× 1½); B3, cyathium (× 3), B2 & B3 from *Richards* 12738. Drawn by
Eleanor Catherine.

Var. **pudibundum** —Carter in Kew Bull. **55**: 435 (2000).

Leaf surfaces glabrous, smooth; margins crisped, with crowded cartilaginous teeth.

Zambia. W: Mwinilunga Distr., near Matonchi Farm, fl. & fr. 18.xi.1962, *Richards* 17276 (K). Also in southeastern Dem. Rep. Congo (Katanga Province). Shallow soil over laterite, in open woodland; c. 1350 m.

Var. **lanatum** S. Carter in Kew Bull. **55**: 436 (2000). Type: Zambia, Zambezi R. Rapids, fl. & fr. 9.xi.1962, *Richards* 16996 (K, holotype).

Leaf upper surface and margins densely covered with cartilaginous hair-like outgrowths 1–1.5 mm long, lower surface glabrous.

Zambia. W: Mwinilunga Distr., Zambezi R. Rapids, c. 6.5 km from Kalene Mission, fl. & fr. 9.xi.1962, *Richards* 16996 (K).
Known only from one collection. Among rocks in gritty soil; c. 1200 m.
Var. *rotundifolium* Malaisse & Lecron (in Bull. Séanc. Acad. Roy. Sc. Outre-Mer **40**: 406 (1994)), with orbicular leaves to 5 cm in diameter, occurs in Dem. Rep. Congo (Katanga Province).

6. **Monadenium pseudoracemosum** P.R.O. Bally in Candollea **17**: 30 (1959); Gen. Mon.: 47 (1961). —S. Carter in F.T.E.A., Euphorbiaceae, part 2: 545 (1988). Type from Tanzania.

Var. **lorifolium** P.R.O. Bally in Candollea **17**: 31 (1950); Gen. Mon.: 49 (1961). Type: Zambia, Kumbula (Nmbulu) Island, fl. & fr. 11.iv.1955, *Richards* 5399 (K, holotype; EA).

Geophyte; rootstock tuberous, 10 × 1.5 cm, cylindrical, producing a woody stem up to 6 cm long below ground. Annual (aerial) stems to 7 cm high. Leaves subsessile, up to 11 × 2 cm, linear-oblanceolate, margin minutely crisped, lower surface often tinged red; stipules minute, glandular. Cymes on peduncles to 2.5 cm long, simple or reduced to solitary cyathia, with cyme branches to 1.5 cm long; bracts joined in a bract-cup c. 6 × 10 mm, ± equal to the involucre, notched to halfway between acute apices, green with darker veining. Cyathia c. 5 × 4 mm, with cup-shaped involucres; glandular rim c. 2 mm high, pink; lobes c. 1.3 × 1.3 mm, rounded, denticulate. Male flowers: bracteoles few, 3 mm long, filamentous; stamens 4 mm long. Female flower: styles 2 mm long, joined to halfway, with thickened bifid apices. Capsule exserted on a pedicel to 6 mm long, c. 5 × 6 mm, obtusely 3-lobed, with a fleshy ridge along each angle. Seeds c. 2.8 × 2.5 mm, subglobose, black with shallow yellowish tubercles; caruncle sessile, 1 mm in diameter.

Zambia. N: Mbala Distr., Kumbula (Nmbulu) Island, fl. & fr. 11.iv.1955, *Richards* 5399 (EA; K).
Known only from the type collection. Lake shore, on steep slopes in dry rocky soil; 770 m.
Var. *pseudoracemosum*, with obovate leaves to 7.5 × 6 cm and margins entire, is known only from southeast Tanzania near the shores of Lake Tanganyika.

7. **Monadenium discoideum** P.R.O. Bally in Candollea **17**: 26 (1959); Gen. Mon.: 88 (1961). Type: Zambia, Mufulira, fl. 3.xii.1948, *Cruse* 435 (K, holotype).

Geophyte; rootstock blackish, tuberous, up to c. 7 × 5 cm, producing a few simple erect stems up to 25 cm high; the whole plant densely covered with short bristly hairs. Leaves subsessile, up to 6.5 × 2.5 cm, oblanceolate, margins entire; stipules glandular, c. 0.5 mm in diameter. Cymes axillary, simple, peduncles c. 6 mm long; bracts joined in a bract-cup, up to 11 × 18 mm, not or only slightly notched, nodding and completely enveloping the involucre, yellowish-green or often flushed reddish. Cyathia c. 4 × 3.5 mm, with barrel-shaped involucres; glandular rim c. 1.5 mm high, yellowish; lobes c. 1 × 1 mm, rounded, denticulate. Male flowers: bracteoles few, c. 2 mm long; stamens c. 3.5 mm long. Female flower: perianth distinctly 3-lobed, 3 mm in diameter; styles 0.75 mm long, with bifid apices. Capsule exserted on a reflexed pedicel c. 5 mm long, c. 4 × 3.5 mm, obtusely lobed, bristly. Seeds 2.25 × 2 mm, subglobose, black, smooth; caruncle 1 mm in diameter, cap-like, shortly stipitate, pale yellow.

Zambia. W: Kitwe Distr., Mwekera, fl. 5.xii.1953, *Fanshawe* 547 (K; NDO).

Also occurs in adjacent southeastern Dem. Rep. Congo (Katanga Province), but restricted overall to an area north of Kitwe. *Brachystegia* woodland; 1200–1250 m.

8. **Monadenium mafingense** Hargr. in Cact. Succ. J. U.S. **53**: 292 (1981); Succ. Spurges Malawi: 61 (1987). Type: Malawi, Mafinga Hills, above Mulembe, fl. 22.xii.1978, *Hargreaves* 616 (MAL, holotype).

Geophyte; rootstock tuberous, up to c. 10 × 5 cm, producing c. 4 stems to 5 cm long, spreading at ground level. Leaves clustered at stem apices, sessile, c. 3.5 × 2.5 cm, obovate, reddish, glabrous to sparsely hairy on the upper surface, with scattered cartilaginous hairs c. 0.75 mm long on the lower surface; margin ± crisped and densely fringed with short hairs. Cymes simple or reduced to solitary cyathia, with peduncles and cyme branches 4–6 mm long; bracts free, 5 × 8 mm, broadly ovate, yellowish-green to red-tinged. Cyathia 6 × 4 mm, with broadly funnel-shaped involucres; glandular rim spreading, c. 1 mm wide, bright red. Male flowers: bracteoles few, c. 2.5 mm long; stamens 3 mm long. Female flower: styles 2.5 mm long, with thickened, bifid apices. Capsule 4 × 3.5 mm, obtusely lobed, with a pair of minute fleshy ridges along the sutures, exserted on a reflexed pedicel c. 6 mm long. Seeds c. 2.75 × 1.25 mm, oblong, 4-angled, densely and minutely tuberculate, pale grey; caruncle 1 mm wide, mushroom-shaped, stipitate, yellow.

Zambia. N: Isoka Distr., Mafinga Mts., fr. 25.v.1973, *Fanshawe* 11936 (K; NDO). **Malawi**. N: Chitipa Distr., Mafinga Mts., fl. & fr. 2.iii.1982, *Brummitt, Polhill & Banda* 16253 (K; MAL).
Known only from the Mafinga Mts. Crevices in faces of rock outcrops, in montane grassland; 1800–2250 m.

9. **Monadenium capitatum** P.R.O. Bally in Candollea **17**: 26 (1959); Gen. Mon.: 34 (1961). —S. Carter in F.T.E.A., Euphorbiaceae, part 2: 546, fig. 102/4–5 (1988); in Kew Bull. **55**: 436 (2000). Type from Tanzania.
    Monadenium fanshawei P.R.O. Bally, Gen. Mon.: 26 (1961). Type: Zambia, 6 km from Kitwe, fl. 2.i.1957, *Fanshawe* 2913 (K, holotype).

Fleshy geophytic herb; with a tuberous tapering root, up to 16 × 6 cm. Stems annual, to 20–50(90) cm high, simple or branched only at the base, minutely scabrid or occasionally glabrous. Leaves subsessile, 9–13 × 2.5–4 cm, oblanceolate to obovate; margin with small distant teeth, sometimes gland-tipped; stipules c. 0.5 mm long, glandular. Cymes axillary, subsessile, 1–2-forked; bracts free, ± equalling the involucres in length, c. 5 × 4 mm, oblong, apex apiculate. Cyathia c. 5 × 3.5 mm, with barrel-shaped involucres; glandular rim c. 1.5 mm high, margin crenulate, white to pink; lobes c. 1.2 × 1.2 mm, rounded, denticulate. Male flowers: bracteoles few, c. 1.75 mm long, filamentous, laciniate; stamens c. 3 mm long. Female flower: styles c. 1 mm long, with thickened bifid apices. Capsule c. 4.5 × 4 mm, acutely 3-lobed, smooth to minutely papillose, exserted on a reflexed pedicel to 5 mm long. Seeds 2.2 × 1.2 mm, oblong, 4-angled, minutely verrucose, pale grey; caruncle c. 1 mm wide, pointed, stipitate, yellow.

Zambia. N: Isoka, fl. 21.xii.1962, *Fanshawe* 7188 (K; NDO). W: Kitwe, fr. 27.v.1967, *Fanshawe* 10087 (K; NDO).
Also in Tanzania. Granite outcrops usually in shade of evergreen thickets, and in high rainfall miombo; 1200–1400 m.

10. **Monadenium herbaceum** Pax in Bot. Jahrb. Syst. **45**: 241 (1910); in F.T.A. **6**, 1: 460 (1911). —Bally, Gen. Mon.: 43 (1961). —Malaisse, Lecron & Schaijes in Bull. Séances Acad. Roy. Sci. Outre-Mer **40**: 404 (1995). Type from Dem. Rep. Congo.

Geophyte with a tuberous subglobose root 2–6 × 2.5–8 cm. Stems annual, single, unbranched, up to 30(110) cm high, scabrid. Leaves subsessile, to 5–7 × 1.5 cm, lanceolate, margin entire, midrib keeled beneath with a fleshy crenulate ridge; stipules c. 0.5 mm long, glandular. Cymes axillary, 1–2-forked, on peduncles 5–20 mm long; cyme branches c. 2–4 mm long; bracts free, slightly longer than the cyathia, c. 5–8 × 5 mm, ovate, acute at the apex, whitish with green veining. Cyathia 4–6 × 3–4 mm, with barrel-shaped involucres; glandular rim 1.5 mm high, crenulate, white;

lobes c. 1.5 × 2 mm, rounded, fimbriate. Male flowers: bracteoles 2 mm long, filamentous; stamens 4 mm long. Female flower: styles 1.5 mm long, with thickened bifid apices. Capsule 4.5–5 × 3.5–4 mm, oblong, deeply lobed, smooth, exserted on a reflexed pedicel 4.5 mm long. Seeds c. 2.75 × 1.25 mm, oblong, 4-angled, densely and minutely verrucose, dark brown; caruncle c. 1.5 mm in diameter, cap-like, shortly stipitate, pale yellow.

Zambia. N: Kaputa Distr., Kundabwika Falls, fr. 17.iv.1989, *Radcliffe-Smith, Pope & Goyder* 5718 (K; LISC).
Also in southeastern Dem. Rep. Congo (Katanga Province). Sandy soil in open *Brachystegia* woodland; c. 1035 m.

11. **Monadenium rugosum** S. Carter in Kew Bull. **55**: 437 (2000). TAB. **87**. Type from Tanzania.

Glabrous fleshy herb, shortly rhizomatous with tuberous roots. Stem unbranched, up to 50 cm high and c. 8 mm in diameter. Leaves fleshy, up to 10 × 7 cm, broadly obovate, tapering into a winged petiole c. 1 cm long, toothed on the margins, distinctly roughened (minutely rugose or verrucose) on the upper surface, midrib keeled on the lower surface; stipules glandular, c. 1 mm wide. Cymes axillary, 1–2-forked; peduncles c. 10 mm long, cyme branches to 4 mm long; bracts c. 7 × 6 mm, ovate, apiculate, joined at the base for 1–1.5 mm, green with darker green or purplish veining. Cyathia 4.5 × 4.5 mm, with cup-shaped involucres; glandular rim 1.5 mm high, white with a red margin; lobes 1 × 1.5 mm, rounded, denticulate. Male flowers: bracteoles few, 2 mm long, laciniate, plumose; stamens 3 mm long. Female flower: perianth a 3-lobed rim; styles 2 mm long, joined for 0.75 mm, apices minutely bifid. Capsule 5.5 × 4.5 mm, oblong, deeply lobed, angles with a pair of minute fleshy ridges. Seeds 3 × 1.25 mm, oblong, 4-angled, verrucose, grey with yellowish warts; caruncle 1.25 mm in diameter, cap-like, stipitate, yellow.

Mozambique. N: Montepuez Distr., Montepuez–Nairoto (Nantulo) road, fl. & fr. 8.iv.1964, *Torre & Paiva* 11753 (LISC).
Also in southeastern Tanzania. *Brachystegia* woodland; 250–500 m.

12. **Monadenium fwambense** N.E. Br. in F.T.A. **6**, 1: 461 (1911). —Bally, Gen. Mon.: 78 (1961). Type: Zambia, Fwambo, fl. ix.1893, *Carson* 17 (K, holotype).

Glabrous geophyte, with an elongated, tuberous root to c. 8 cm in diameter. Stems unbranched, up to 20 cm high. Leaves subsessile, up to 12 × 0.6 cm and narrowly lanceolate, or up to 4 × 1.5 cm and obovate, midrib prominent beneath. Cymes axillary, simple; peduncles c. 3 mm long, longitudinally ridged; bracts joined in a bract-cup c. 6.5 × 4.5 mm, overtopping the involucre, deeply notched sometimes almost to the base, between very acute lobes with keeled midribs, green or flushed pink. Cyathia c. 4 × 3.5 mm, with cup-shaped involucres; glandular rim 1.5 mm high, white; lobes c. 1 × 1 mm, rounded, denticulate. Male flowers: bracteoles few, 2 mm long, laciniate; stamens c. 3 mm long. Female flower: styles 2 mm long, joined for c. 1 mm, and minutely bifid at the apex. Capsule 4 × 3.5 mm, ± oblong, obtusely lobed, ridged along the angles. Seeds 2.25 × 1.25 mm, oblong, 4-angled, minutely verrucose, grey; caruncle 1 mm in diameter, cap-like, shortly stipitate, yellow.

Zambia. N: Chinsali Distr., Mbwingimfumu Hills, fl. 4.ii.1955, *Fanshawe* 1985 (K; NDO); Kaputa Distr., Mulimbi–Muchinga Escarpment, c. 28 km north of Mporokoso, fl. & fr. 15.iv.1989, *Radcliffe-Smith, Pope & Goyder* 5699 (K).
Not known elsewhere. Rock cervices in granite outcrops, among rocks on stony hillsides and on laterite; 1500–1870 m.

13. **Monadenium crenatum** N.E. Br. in F.T.A. **6**, 1: 461 (1911). —Bally, Gen. Mon.: 77 (1961). Type: Mozambique, Lion's Creek, fl. 8.iv.1898, *Schlechter* 12204 (K, holotype).

Glabrous herb, probably geophytic with a tuberous root. Stem unbranched, to 12.5 cm high, glabrous. Leaves sessile, c. 6 × 1 cm, narrowly lanceolate, margins crisped; stipules glandular, c. 0.5 mm in diameter. Cymes axillary, 1–2-forked; peduncles and cyme branches to 1 cm long; bracts joined in a bract-cup, nodding, c.

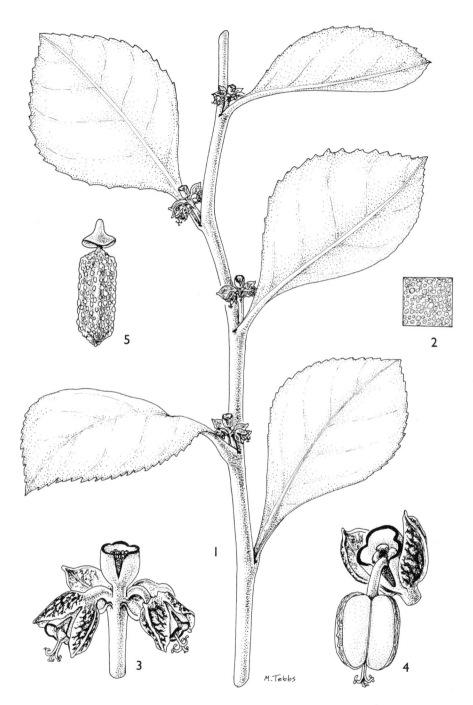

Tab. 87. MONADENIUM RUGOSUM. 1, portion of flowering branch (× ⅔); 2, portion of leaf showing rough surface (2 ⅔); 3, flowering cyme (× 2), 1–3 from *Bidgood, Abdallah & Vollesen* 1976; 4, fruiting cyathium (× 4); 5, seed showing cap-like caruncle (× 8), 4 & 5 from *Torre & Paiva* 11753. Drawn by Margaret Tebbs. From Kew Bull.

6 × 5 mm, notched to halfway or almost to the base between acute lobes. Cyathia 4 × 3.5 mm, with cup-shaped involucres; glandular rim crenulate, 1.5 mm high; lobes 1 × 1.5 mm, rounded, denticulate. Male flowers: bracteoles free, 1.5 mm long, laciniate; stamens c. 2.5 mm long. Female flower: styles 1.5 mm long, joined to halfway and minutely bifid at the apex. Capsule and seeds not seen.

Mozambique. MS: Gondola Distr., Lion's Creek, between Mutare (Umtali) and Fontesvilla, fl. 8.iv.1898, *Schlechter* 12204 (K).
Known only from the type specimen. Habitat not stated; c. 304 m.

14. **Monadenium parviflorum** N.E. Br. in F.T.A. **6**, 1: 458 (1911). —Bally, Gen. Mon.: 53 (1961). —Binns, First Check List Herb. Fl. Malawi: 51 (1968). —Hargreaves, Succ. Spurges Malawi: 67 (1987). —S. Carter in Kew Bull. **55**: 439 (2000). Type: Malawi, Chitipa (Fort Hill), fl. & fr. 21.vii.1896, *Whyte* s.n. (K, holotype).
Monadenium laeve forma *depauperata* P.R.O. Bally in Candollea **17**: 35 (1959); Gen. Mon.: 85 (1961). Type from Tanzania.
Monadenium depauperatum (P.R.O. Bally) S. Carter in Kew Bull. **42**: 908 (1987); in F.T.E.A., Euphorbiaceae, part 2: 548 (1988).
Monadenium hirsutum sensu Hargreaves, Succ. Spurges Malawi: 69 (1987), non P.R.O. Bally excl. desc.

Perennial herb to 30(60) cm high, probably geophytic, with a subglobose tuberous root up to c. 3 cm in diameter. Stem usually unbranched, sparsely hairy. Leaves 5.5–10 × 1–4.5 cm, oblanceolate, tapering into a petiole up to 10 mm long; lamina with scattered ± cartilaginous hairs to 1 mm long on both surfaces, the hairs more numerous on the margins; midrib on the lower surface with a wing-like keel 0.5–0.75 mm wide; stipules glandular. Cymes terminal and axillary, crowded towards the stem apex, 2–4-forked; peduncle up to 2 cm long; cyme branches c. 4 mm long, longitudinally ridged and sometimes with cartilaginous outgrowths below the bracts; bracts joined in a bract-cup c. 4 × 6.5 mm, notched to nearly halfway between rounded lobes, enveloping the involucre, pale green or pinkish with slightly darker veining. Cyathia c. 3.5 × 2.5 mm, with cup-shaped involucres; glandular rim c. 1.5 mm high, creamy-white; lobes c. 0.75 × 0.75 mm, denticulate. Male flowers: bracteoles few, 1.5 mm long, filamentous; stamens 3 mm long. Female flower: perianth a 3-lobed rim, 1.5 mm in diameter; styles 1.25 mm long, joined to halfway and shortly bifid at the apex. Capsule c. 4 × 3 mm, acutely lobed, smooth, with fleshy ridges along the sutures, exserted on a reflexed pedicel c. 4 mm long. Seeds c. 2.2 × 1 mm, oblong, 4-angled, grey with minute whitish warts; caruncle cap-like, 0.75 mm in diameter, subsessile.

Zambia. N: Mbala Distr., Kawimbe, Nachalanga (Nchalanga) Hill, fl. & fr. 13.iv.1961, *Richards* 15038 (K). **Malawi**. N: Chitipa Distr., Kaseye Mission, 16 km east of Chitipa, fl. & fr. 25.iv.1977, *Pawek* 12647 (K; MAL); Chambo Mission, 4 km north of Chisenga, fl. 3.iii.1982, *Brummitt, Polhill & Banda* 16305 (K).
Also in southwestern Tanzania. Rock crevices and on decomposing granite, in high rainfall *Brachystegia* woodland and thickets, on hillsides and on termitaria; 548–1840 m.
This species exhibits much variation in size of the plant, leaves and involucral bracts, apparently in response to environmental conditions.

15. **Monadenium hirsutum** P.R.O. Bally in Candollea **17**: 27 (1959); Gen. Mon.: 86 (1961). TAB. **86**, fig. B. Type: Zambia, Kundabwika Falls, fl. & fr. v.1931, *E.G. Walter* 1 (K, holotype).

Geophyte, up to 25 cm high, with a tuberous root c. 4 × 3.5 cm. Stem with scattered, slender hairs. Leaves subsessile, to 8 × 1 cm, lanceolate to linear-lanceolate, margins crisped, midrib prominent below, both surfaces densely covered with filamentous hairs to 1 mm long; stipules glandular, minute. Cymes terminal and axillary, 3–5-forked; peduncles to 7 cm long; cyme branches to 1 cm long; bracts joined to form a bract-cup up to 8.5 × 10 mm, shortly notched with rounded lobes, midribs prominent, margins undulate, enveloping the involucre. Cyathia c. 4 × 4 mm, with cup-shaped involucres; glandular rim 1 mm high, white with a red margin; lobes c. 1 × 1 mm, denticulate. Male flowers: bracteoles few, 2 mm long; stamens 3.5 mm long. Female flower: styles 2 mm long, joined to halfway and deeply bifid at the apex. Capsule c. 4.5 × 4 mm, 3-lobed, with minute fleshy ridges along the sutures, exserted on a pedicel c. 4 mm long. Seeds c. 2.5 × 1.4 mm, oblong, 4-angled, densely and very minutely verrucose, pale brown; caruncle cap-like, 1 mm in diameter, shortly stipitate.

Zambia. N: Kaputa Distr., Mweru Wantipa, on road to Kanjiri (Kangiri), fl. & fr. 6.iv.1957, *Richards* 9060 (K; LISC); Yendwe (Iyendwe) Valley, Lufubu R., fl. & fr. 12.iv.1967, *Richards* 22204 (K).
Not known elsewhere. On rocks and amongst short grasses in woodland; 900–1035 m.
Bally erroneously located Kundabwika Falls as lying east of Lusaka. *Burtt* 6434 and *Richards* 5293 cited by him in "The Genus Monadenium" are specimens of *M. parviflorum*.

16. **Monadenium laeve** Stapf in Hooker's Icon. Pl. **27**: t. 2666 (1900). —N.E. Brown in F.T.A. **6**, 1: 456 (1911). —Bally, Gen. Mon.: 83 (1961). —Binns, First Check List Herb. Fl. Malawi: 51 (1968). —Hargreaves, Succ. Spurges Malawi: 70 (1987). Type: Malawi, between Kondowe and Karonga, fl. & fr. viii.1896, *Whyte* s.n. (K, holotype, two sheets).

Perennial herb, with a woody rhizomatous root c. 1.5 cm in diameter. Stems glabrous, usually simple, up to c. 80 cm high, often decumbent and then rooting in leaf litter. Leaves to 16 × 7 cm, obovate, tapering at the base into a winged petiole to c. 1 cm long, midrib keeled on the lower surface, the upper or both surfaces with stiff, scattered hairs; stipules glandular, 0.5 mm in diameter. Cymes axillary, 2–3-forked; peduncles glabrous, up to 10 cm long, cyme branches 0.5–2 cm long; bracts joined in a bract-cup to c. 14 × 20 mm, nodding and enveloping the cyathium, notched to nearly halfway between rounded lobes, creamy-white with green veining, midribs prominent. Cyathia c. 4 × 5 mm, with cup-shaped involucres; glandular rim c. 1.25 mm high, white; lobes 1.5 × 1.5 mm, rounded, denticulate. Male flowers: bracteoles few, c. 2.5 mm long, filamentous; stamens c. 4 mm long. Female flower: styles 1.5 mm long, joined to halfway, deeply bifid at the apex. Capsule 5.5 × 5 mm, acutely lobed, angles ridged, smooth, exserted on a reflexed pedicel c. 6 mm long. Seeds 2.5 × 1.25 mm, oblong, 4-angled, minutely verrucose, grey with pale brown warts; caruncle 1 mm in diameter, cap-like, shortly stipitate, yellow.

Malawi. N: Mzimba Distr., 9.5 km north of Mzambazi, fl. 10.iii.1978, *Pawek* 13991 (K; MAL; MO); Karonga Distr., by Sere R., below Kayelekera, 30 km WSW of Karonga, fl. & fr. 7.vi.1989, *Brummitt* 18415 (K).
Also in southern Tanzania. On rocky hillsides in evergreen forest and *Brachystegia* woodland; 800–1200 m.

17. **Monadenium friesii** N.E. Br. in R.E. Fries, Wiss. Ergebn. Schwed. Rhod.-Kongo-Exped. **1**: 115 (1914). Type: Zambia, Kabwe (Broken Hill), fl. & fr. 4.viii.1911, *Fries* 224 (K, holotype).

Perennial herb, with a subglobose tuberous root to c. 5 cm in diameter. Stems to 45 cm high, unbranched. Leaves to 9 × 6 cm, suborbicular to ovate, sometimes reddened, both surfaces scabridulous with usually numerous stiff cartilaginous hairs, margin tightly crisped and densely hairy, midrib prominent beneath; petiole c. 6 mm long; stipules c. 0.5 mm in diameter, glandular. Cymes axillary, 3–4-forked; peduncles to c. 8 cm long, cyme branches to 15 mm long, both with conspicuous scattered cartilaginous prickles 0.2–1 mm long; bracts joined in a bract-cup enveloping the cyathium, c. 10 × 16 mm, shallowly notched between rounded lobes, whitish with dark green to purplish veining and the surface with scattered cartilaginous outgrowths. Cyathia c. 5 × 4 mm, with cup-shaped involucres; glandular rim 1.25 mm high, white; lobes 1 × 1 mm, rounded, denticulate. Male flowers: bracteoles few, 2 mm long; stamens 3 mm long. Female flower: ovary covered with cartilaginous hairs 0.2–0.5 mm long; styles 2 mm long, joined to nearly halfway, shortly bifid at the apex. Capsule 5 × 4 mm, obtusely lobed, surface cartilaginous and wrinkled, exserted on a reflexed pedicel c. 5 mm long. Seeds 3 × 1.25 mm, oblong, 4-angled, minutely verrucose, pale brown; caruncle cap-like, 1 mm in diameter, stipitate, yellow.

Zambia. N: Isoka Distr., 5 km east of Kampumbo School, on road between Isoka and

Muyombe, fl. & fr. 15.iv.1986, *Philcox, Pope & Chisumpa* 9932 (K). C: Luangwa Distr., Katondwe, fr. 20.vi.1966, *Fanshawe* 9726 (K; NDO). E: Katete, fl. & fr. 24.iii.1955, *Exell, Mendonça & Wild* 1149 (BM; LISC).
Not known elsewhere. On rocks and rocky hillsides in *Brachystegia* woodland; 375–1200 m.

18. **Monadenium torrei** L.C. Leach in Garcia de Orta, Sér. Bot. **1**: 37, fig. 3 & t. v (1973). —S. Carter in F.T.E.A., Euphorbiaceae, part 2: 551 (1988). TAB. **88**, fig. A. Type: Mozambique, Montepuez, fl. 9.iv.1964, *Torre & Paiva* 11790 (LISC, holotype; COI; K; LMU; PRE; SRGH).
Monadenium sp. aff. spinescens sensu P.R.O. Bally, Gen. Mon.: 101 (1961).

Sturdy shrub, erect, up to 3 m high, glabrous. Stem sparsely branched, with smooth bark. Branches c. 1.5 cm in diameter, with shallow tubercles below the leaf scars in 5 longitudinal series, each crowned by a stout curved spine up to 8 mm long. Leaves crowded at branch tips, sessile, up to 12 × 3.5 cm, oblanceolate, irregularly toothed on the margins, bright green, midrib keeled on lower surface; stipules 1 mm long, weakly spiny. Cymes 3–6-forked; peduncles up to 10 cm long, longitudinally ridged; cyme branches to 5 cm long, pubescent especially when young; bracts with prominent midribs and shortly pubescent towards the base, yellowish-green, joined in a bract-cup c. 1 × 1.5 cm, notched almost to the base between rounded apices. Cyathia c. 10 × 5 mm, with barrel-shaped involucres, pubescent; glandular rim to 6 mm high, yellow; lobes c. 5 × 2 mm, oblong, deeply toothed, pubescent. Male flowers: bracteoles few, filamentous, plumose, c. 4 mm long; stamens 7 mm long. Female flower: styles 3 mm long, joined at the base, deeply bifid at the apex. Capsule c. 5.5 × 4.5 mm, obtusely lobed, pubescent at the base, exserted on a pedicel to c. 4 mm long. Seeds 3 × 1.5 mm, oblong, 4-angled, densely verrucose, pale grey; caruncle cap-like, 1 mm in diameter, stipitate.

Mozambique. N: Mt. Mecótia (Mt. M'kota), st. 1907, *Stocks* 146 (K).
Also in southeastern Tanzania. Among rocks in deciduous woodland; 300–560 m.

19. **Monadenium lugardiae** N.E. Br. in Bull. Misc. Inform., Kew **1909**: 138 (1909); in F.T.A. **6**, 1: 452 (1911). —R.A. Dyer in Fl. Pl. South Africa **6**: 223 (1926). —Burtt Davy, Fl. Pl. Ferns Transvaal: 290 (1932). —Bremekamp & Obermeyer [Scientific Results of the Vernay-Lang Kalahari Expedition, March to September, 1930], in Ann. Transvaal Mus. **16**: 421 (1935). —White, Dyer & Sloane, Succ. Euphorb. **2**: 943 (1941). —Suessenguth & Merxmüller, [Contrib. Fl. Marandellas Distr.] Proc. & Trans. Rhod. Sci. Ass **43**: 84 (1951). —Bally, Gen. Mon.: 56 (1961), —Court, Succ. Fl. South. Africa: 29 (1981). —Hargreaves, Succ. Spurges Malawi: 72 (1987); Succ. Botswana: 9 (1990). TAB. **88**, fig. B. Type: Botswana, Khwebe (Kwebe) Hills, fl. & fr. 31.viii.1897, *Mrs. E.J. Lugard* 22 (K, holotype).

Perennial with thick fleshy roots. Stems succulent, glabrous, branching from the base, 10–60 cm high, erect or shortly decumbent and rooting, 1.5–3 cm in diameter, cylindrical, marked with ± rhomboid tessellations (with a diamond-shaped pattern of flattened tubercles); tubercles c. 1.5 × 1 cm, with circular leaf scars at the apices; leaf scars 2 mm in diameter with a narrow ± horny rim around the base. Leaves subsessile, crowded towards the stem and branch apices, to 9 × 4 cm, obovate, fleshy, minutely puberulous, margins often ± crisped, sometimes flushed reddish; stipules on young growth modified into a cluster of 3–5 soft spines 0.5–2 mm long which soon shrivel. Cymes axillary, simple; peduncles 5–8 mm long; cyme branches 2–4 mm long, minutely puberulous; bracts joined in a bract-cup c. 7 × 7 mm, shortly notched between acute lobes and prominent midribs, minutely puberulous, often tinged pinkish. Cyathia c. 4 × 3.5 mm, with cup-shaped involucres; gland c. 2 mm high, cream with yellow rim; lobes 1 × 1.5 mm, rounded, denticulate. Male flowers: bracteoles 2.5 mm long, laciniate; stamens 4 mm long. Female flower: perianth 3-lobed, 2.5 mm in diameter; styles 1.5 mm long, joined at the base, deeply bifid at the apex. Capsule c. 6 × 6 mm, obtusely lobed, with a pair of fleshy crested ridges along the sutures, exserted on a reflexed pedicel to 8 mm long. Seeds c. 3.75 × 1.75 mm, oblong, 4-angled, minutely and shallowly verrucose, pale brownish-grey; caruncle cap-shaped, 1 mm in diameter, shortly stipitate.

Botswana. N: c. 19 km south of border on Francistown road, fl. 14.i.1960, *Leach & Noel* 24 (K; SRGH). **Zimbabwe.** N: Binga Distr., 8 km southeast of Mwenda Research Station, fr. 7.vi.1966, *Grosvenor* 131 (K; SRGH). W: Hwange Distr., 9.5 km east of Kamativi, fr. 1956, *Leach* 5192 (K;

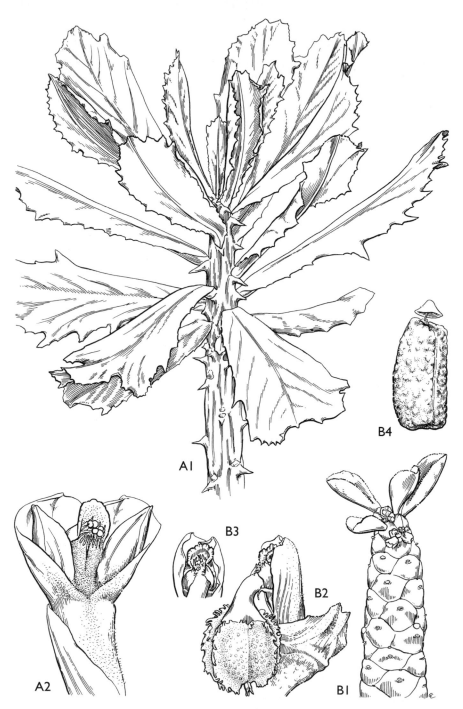

Tab. 88. A. —MONADENIUM TORREI. A1, leafy branch (× ²⁄₃); A2, cyathium (× 3), A1 & A2 from *Torre & Paiva* 11790. B. —MONADENIUM LUGARDIAE. B1, flowering branch (× ²⁄₃); B2, fruiting cyathium (× 3); B3, cyathium, front view (× 3), B1–B3 from *Leach* 5192; B4, seed showing cap-shaped caruncle (× 8), from *Grosvenor* 131. Drawn by Eleanor Catherine.

SRGH). E: 22.5 km south of Mutare (Umtali), fl. & fr. 20.xi.1960, *Leach & Chase* 10495 (COI; K; LISC; SRGH). S: Chivi Distr., 48 km north of Runde (Lundi) R. on Masvingo–Beitbridge road, fl. & fr. 26.ix.1963, *Leach* 11733 (K; LISC; SRGH). **Mozambique**. GI: Chicualacuala Distr., 13 km southeast of Pafuri, fl. 11.vii.1964, *Leach & Mockford* 12301 (LISC; SRGH). M: Moamba, fr. 1.xii.1942, *Mendonça* 1530 (LISC).

Also in South Africa (Northern Prov. to KwaZulu-Natal). Granite outcrops, in sandy soil among rocks in open *Brachystegia* woodland and wooded grasslands, and in shade on termite mounds; often in colonies; 115–1100 m.

20. **Monadenium spinulosum** S. Carter in Kew Bull. **55**: 440 (2000). Type: Zimbabwe, Chesa Purchase Land (Reserve), east of Mount Darwin, fl. & fr. 30.iii.1960, *Leach* 9831 (K, holotype; SRGH).

Monadenium schubei sensu White, Dyer & Sloane, Succ. Euphorb. **2**: 947, figs. 1080–1083 (1941). —sensu R.A. Dyer in Fl. Pl. South Africa **28**: t. 1087 (1950). —sensu Bally, Gen. Mon.: 65 (1961), pro parte as to "S. Rhodesia" distribution, non (Pax) N.E. Br.

Monadenium lugardiae sensu Hargreaves, Succ. Spurges Malawi: 72 (1987), as to *Binns* ex *Burdett & Fletcher* 610.

Perennial with thick fleshy roots. Stems succulent, glabrous, sparingly branched, erect to c. 45 cm high or decumbent to 80 cm long and rooting, up to 3 cm in diameter, cylindrical, covered with tubercles in a diamond-shaped pattern (tessellations); tubercles c. 10 × 8 mm and up c. 5 mm high, crowned with circular leaf scars 3 mm in diameter. Leaves subsessile, crowded towards the stem apex, up to 10 × 4.5 cm, obovate, tapering to a narrow base, fleshy, minutely puberulous, margins crisped, midrib prominent beneath; stipules modified into a cluster of 5–6 firm prickles, 1–3 mm long, arranged around the base of the leaf scars. Cymes axillary, simple or occasionally 2-forked; peduncles to 8 mm long; cyme branches c. 3 mm long; bracts joined in a bract-cup c. 6 × 9 mm, shortly notched between rounded lobes, sometimes minutely puberulous, yellowish-green. Cyathia c. 6 × 5 mm, with cup-shaped involucres; glandular rim c. 2.5 mm high, minutely crenulate, cream with yellow; lobes 1.25 × 2 mm, rounded, denticulate. Male flowers: bracteoles 2 mm long, filamentous; stamens 4 mm long. Female flower: perianth 3-lobed, 3.5 mm in diameter; styles 2 mm long, joined at the base, deeply bifid at the apex. Capsule c. 8 × 8 mm, obtusely lobed, with a pair of fleshy crested ridges 0.5 mm wide along the sutures, exserted on a reflexed pedicel to 8 mm long. Seeds c. 4 × 1.8 mm, oblong, 4-angled, minutely and shallowly verrucose, pale brownish-grey; caruncle cap-shaped, 1 mm in diameter, shortly stipitate.

Zimbabwe. N: Mutoko (Mtoko), fl. & fr. iii.1945, *Christian* in *PRE* 27326 (K; PRE). **Malawi**. S: Nsanje Distr., Lower Shire, cult. st. iii.1971, *Fletcher & Burdett* in *Binns* 610 (MAL). **Mozambique**. T: Magoe Distr., Mphende (Màgué), 5 km from Cahora Bassa (Cahorabassa) Dam site, fl. & fr. 19.ii.1968, *Torre & Correia* 17736 (LISC). MS: Tambara Distr., Chemba, 22 km from Nhacolo (Tambara), fr. 14.v.1971, *Torre & Correia* 18426 (LISC).

Not known elsewhere. Granite outcrops and amongst rocks on hillsides, in *Brachystegia* woodland, in shade; 65–1200 m.

62. PEDILANTHUS Neck. ex A. Poit.

Pedilanthus Neck. ex A. Poit. in Ann. Mus. Natl. Hist. Nat. **19**: 388 (1812), *nom. conserv.* —Boissier in de Candolle, Prodr. **15**, 2: 4 (1862). —Dressler in Contrib. Gray Herb., No. 182: 97 (1957). —S. Carter in F.T.E.A., Euphorbiaceae, part 2: 564 (1988). —Radcliffe-Smith, Gen. Euphorbiacearum: 418 (2001).

Shrubs or small trees with woody or fleshy branches and a milky latex, monoecious. Leaves shortly petiolate, entire, with small stipules. Inflorescence with cyathia in dichotomous axillary or terminal cymes; bracts paired, persistent. Involucres with 5 unequal lobes and 2, 4 or 6 glands enclosed within an adaxial spur-like extension of the involucre formed from 4 gland appendages, often brightly coloured. Male flowers in 5 groups, usually bracteolate. Female flower pedicellate, with the perianth reduced to a rim below the ovary; ovary 3-locular, with 1 pendulous ovule in each locule; styles 3, connate, with bifid stigmas. Fruit 3-lobed, usually a dehiscent capsule, occasionally indehiscent. Seeds smooth or tuberculate, without a caruncle.

A genus of 16 species in Central America (mainly in Mexico), northern South America and the West Indies. A few species, particularly *Pedilanthus tithymaloides*, are cultivated in the tropics.

Pedilanthus tithymaloides (L.) A. Poit. in Ann. Mus. Natl. Hist. Nat. Paris **19**: 390 (1812). Type from Central America.

Subsp. **smallii** (Millsp.) Dressler in Contrib. Gray Herb., No. 182: 152 (1957). Type from southern North America (Florida).
    Pedilanthus smallii Millsp. in Field Mus. Pub. Bot. **2**: 358 (1913).

Succulent shrub up to 3 m high; branches distinctly zigzag. Leaves up to 7 × 3.5 cm, broadly ovate to lanceolate, fleshy, slightly glaucous, sometimes variegated with yellowish-green or pink which can cause puckering, quickly deciduous. Cymes in terminal and axillary clusters; cyathia inconspicuous, with 4 glands enclosed by the pink or red beak-shaped involucral extension c. 15 mm long.

Malawi. C: Salima Distr., Lake Nyasa Hotel, fl. 27.vii.1951, *Chase* 3847 (K; SRGH). **Mozambique**. N: Mocímboa da Praia, Mitamba (M'tamba), fl. 14.x.1906, *Stocks* 51 (K). Z: Nicoadala Distr., Quelimane (Quillimane), fl. xi.1862, *Kirk* s.n. (K). M: Maputo, Jorge Dimitrov (Benfica), fl. 25.x.1989, *Groenendijk* 2211 (K; LMU).
    Cultivated as an ornamental in gardens and often grown as a hedge, mostly near the coast; 0–480 m.
    The species includes a number of subspecies, but subsp. *smallii* is the one most commonly cultivated. Plants have occasionally become naturalised in India, but apparently not, so far as is known, in tropical Africa.

INDEX TO BOTANICAL NAMES